Nanobiosensors for Agricultural, Medical and Environmental Applications

Mohd. Mohsin • Ruphi Naz • Altaf Ahmad
Editors

Nanobiosensors for Agricultural, Medical and Environmental Applications

Springer

Editors
Mohd. Mohsin
Department of Biosciences
Jamia Millia Islamia
New Delhi, Delhi, India

Ruphi Naz
Department of Biosciences
Jamia Millia Islamia
New Delhi, Delhi, India

Altaf Ahmad
Department of Botany
Aligarh Muslim University
Aligarh, Uttar Pradesh, India

ISBN 978-981-15-8345-2 ISBN 978-981-15-8346-9 (eBook)
https://doi.org/10.1007/978-981-15-8346-9

This Springer imprint is published by the registered company Springer Nature Singapore Pte Ltd.
The registered company address is: 152 Beach Road, #21-01/04 Gateway East, Singapore 189721, Singapore

Preface

Nanotechnology has unlocked the frontier area of research for rapid advancement in the science and technology. It continuously provides tremendous opportunities in agriculture, medicine, food, environment, and electronics. One of the significant areas of the advancements in nanotechnology is the development of nanobiosensors. Molecular sensing and molecular dynamics are the key areas of nanobiotechnology that utilize various approaches like changes in conformation, optical absorbance, charge distribution, emission, electrical conductivity across a single or interconnected molecule, generated as a reaction to a target input. Nanobiotechnology-based biosensors are drawing the attention of scientists towards its broad-spectrum use. Biosensor is a device that can read changes in the physical stimulus and convert them into recordable signals.

Nanobiosensors are portable and sensitive devices of chemical and biological agents. The prefix term "Nano" is considered as billionths of a meter. Sensors are developed by incorporating a biological sensing element either intimately connected to or integrated within a transducer. The sensor actively responds with the binding of a single molecule of particular metabolites, nucleic acid, or protein within the living system. Binding of the ligand to the sensor surface depends mainly on the surface chemistry. Nanobiosensor considers the sensitivity, specificity, and dynamic range of the sensor for enhanced functionality. The sensitivity is described as the total output signal generated per analyte concentration or the minimum quantity of the analyte resolved by the sensor over a background signal, better called a limit of determination (LOD). In the present era, nanobiosensor technology has revolutionized all domains of science as the health care industry, food industry, improvement in environmental quality, crop productivity, agricultural, pharmaceuticals, and metabolic complications. Development of interference-free endotoxin detection systems, paper inkjet sensor, immunosensor, enzyme sensor, microfluidic sensor devices, real-time aptamer-based sensor and ultrasensitive biosensors for the detection of the virus, attract the scientific mind towards a particular field. Further, nanoparticles of gold, magnetic, silver, carbon nanotubes, and quantum dots have been strongly exploring their application as biosensors.

With the emergence of the fluorescence protein, fluorescence-based biosensor has become potential tools for identifying and monitoring cellular physiology, molecular dynamics, and cell–cell interactions/communication and studying the metabolic pathways inside the cell.

Although researchers actively engaged in developing biosensors for various purposes in different fields, the opportunity has arrived to bring this innovation to the cutting edge, and make it accessible commercially. In this way, there is a need to enhance the technology by the scientists' concern around the world. Therefore, endeavors and assets should be prepared to make biosensors for a vast scope to profit and be useful to the overall population.

In this particular book, we have covered various types of nanobiosensors for applications in the fields of agriculture, environment, and medicine. The book chapters contain a tremendous amount of relevant information that has become available in the context of nanobiosensors. Hopefully, this book will be an excellent source to provide sufficient information to the readers to understand, what are nanobiosensors, how it works, and their application in all related domains.

This book deals with a series of chapters related to nanobiosensors for agricultural, medical, and environmental applications.

As editors, responsibility comes in our account for making the selection of topics covered in this book. Every chapter has been reviewed by esteemed reviewers. We feel great pleasure to work with the authors and co-authors who have submitted their chapters according to the schedule, giving positive responses to the reviewers' comments during revision.

We appreciate the numerous reviewers for accepting our request to look after the chapters and make useful comments to the authors.

New Delhi, India — Mohd. Mohsin
New Delhi, India — Ruphi Naz
Aligarh, India — Altaf Ahmad

Contents

Editors and Contributors

About the Editors

Mohd. Mohsin is an Assistant Professor in Jamia Millia Islamia, New Delhi, India since 2014. He completed his Master's degree in Biosciences from Jamia Millia Islamia, New Delhi, and a Ph.D. degree from Hamdard University, New Delhi. In 2009, he qualified the National Eligibility Test of Council of Scientific and Industrial Research (CSIR) and Junior Research Fellowships of Indian Council of Medical Research. He has been working on Nanobiotechnology & molecular sensors. His research interest focuses on developing "Genetically Encoded FRET-based nanosensors" for healthcare and cellular environmental monitoring. He has presented his research work in various national and international conferences and published many research papers in the journals of international repute. He received a high Impact Factor (>10) publication award in 2014. His research is funded by various agencies of the Govt. of India, SERB, DBT, ICMR, and UGC. He is mentoring several Ph.D. students and postdoctoral fellows. Also, his contribution was considered impactful by the scientific community and was conferred Young Scientist award in 2015 from Govt. of India and also an innovative research award in 2017. He is also serving as an editorial member and reviewer of several journals.

Ruphi Naz is a National Postdoctoral Fellow of Science and Engineering Research Board (NPDF-SERB) at Jamia Millia Islamia, New Delhi. She earned her M.Phil and Ph.D. in Plant Biotechnology at the Department of Botany of Aligarh Muslim University (Aligarh). She has awarded UGC-BSR Fellowship during her Ph.D. She has intense background knowledge of micropropagation and plant transformation and hands-on experience of plant molecular biology techniques as well as the development of FRET-based nanosensor. She has published several research papers in the Journal of International repute. Based on her research work, she has been awarded Dr. DS Kothari Postdoctoral fellowship of the University Grant Commission-UGC (2015–2018) and Senior Research Associate of Council of Scientific and Industrial Research (2019–2020) at the Department of Botany of Aligarh Muslim University, Aligarh, India. She is a life time member of the National and International Scientific Societies. She has presented her research work in a number of conferences nationally and internationally organized in various places in India.

Altaf Ahmad is a Professor of Plant Biotechnology in the Department of Botany, Aligarh Muslim University, Aligarh, India. Dr. Ahmad has a research experience of 22 years and has to his credit more than 125 significant publications. A regular reviewer and on the panel of editorial boards of National and International journals, Dr. Ahmad is also a member of prestigious National and International scientific societies. He has guided thesis research of 30 Ph.D. students (12 as supervisor and 18 as co-supervisor) and over 30 M.Sc. students. He has organized several national and international seminars/symposia. Dr. Ahmad handled several research projects funded by various agencies like BBSRS-UK, DST, DBT, CSIR, UGC, CCRUM, ICAR, and INSA. He was a Visiting Scientist at the Virginia Polytechnic and State University, USA University of Freiburg, University of Aberdeen, and the University of Cologne. Based on his research contribution, Dr. Ahmad has been honored with the INSA Medal for Young Scientist (2003) and Professor LSS Kumar Memorial Award (2003) of Indian National Science Academy, Young Scientist Award (2005) of Council of Science & Technology, Govt. of Uttar Pradesh, and elected as Associate of National Academy of Agricultural Sciences.

Contributors

Altaf Ahmad Department of Botany, Aligarh Muslim University, Aligarh, India

Mohammad Ahmad Department of Physics, Syracuse University, New York, NY, USA

Razi Ahmad Department of Biosciences, Jamia Millia Islamia, New Delhi, India

Akrema Department of Chemistry, Middle East Technical University, Ankara, Turkey

Ghazala Ambrin Department of Botany, Aligarh Muslim University, Aligarh, India

Amreen Department of Botany, Aligarh Muslim University, Aligarh, India

Rizwan Arif Department of Biosciences, Jamia Millia Islamia, New Delhi, India

Tenzin Choephel University Institute of Biotechnology, Chandigarh University, Mohali, India

Shahla Faizan Environmental Physiology Laboratory, Department of Botany, Aligarh Muslim University, Aligarh, India

Saher Fatima Department of Agricultural Microbiology, Aligarh Muslim University, Aligarh, India

Urooj Fatima Department of Botany, Aligarh Muslim University, Aligarh, India

Priyanka Gulia Centre for Nanoscience and Nanotechnology, Jamia Millia Islamia, New Delhi, India

Habiba Kausar Department of Botany, Aligarh Muslim University, Aligarh, India

Summaiyya Khan Department of Biosciences, Jamia Millia Islamia, New Delhi, India

Samrah Manzoor Centre for Nanoscience and Nanotechnology, Jamia Millia Islamia, New Delhi, India

Prabhash Mishra Center for Photonics and 2D Materials, Moscow Institute of Physics and Technology (MIPT), Dolgoprudny, Russia

Mohd. Mohsin Department of Biosciences, Jamia Millia Islamia, New Delhi, India

Zeenat Mushtaq Environmental Physiology Laboratory, Department of Botany, Aligarh Muslim University, Aligarh, India

Rahila Nazir Department of Botany, Jamia Hamdard, New Delhi, India

Ruphi Naz Department of Biosciences, Jamia Millia Islamia, New Delhi, Delhi, India

Neelam P. Negi University Institute of Biotechnology, Chandigarh University, Mohali, India

Manju Nehra Department of Food Science and Technology, Chaudhary Devi Lal University, Sirsa, India

Rubia Noori Department of Biosciences, Jamia Millia Islamia, New Delhi, India

Rahisuddin Department of Biosciences, Jamia Millia Islamia, New Delhi, India

Meryam Sardar Department of Biosciences, Jamia Millia Islamia, New Delhi, India

Tariq Omar Siddiqi Department of Botany, Jamia Hamdard, New Delhi, India

Neha Soleja Department of Biosciences, Jamia Millia Islamia, New Delhi, India

Mohammad Talib Centre for Nanoscience and Nanotechnology, Jamia Millia Islamia, New Delhi, India

Nishant Tripathi Department of Nanoengineering, Samara National Research University, Samara, Russia

Opportunities for Real-Time Monitoring of Biomolecules Using FRET-Based Nanosensors

1

Neha Soleja and Mohd. Mohsin

Abstract

Fluorescence technology has flourished over the past decade. The discovery of green fluorescent protein (GFP) and its variants was a major breakthrough because of its non-invasive nature. GFP, which is responsible for showing the fluorescence, was extracted from the jellyfish, *Aequorea victoria*. Further mutations in this wild-type GFP yielded variants that were brighter, more photostable, and emitted a wide range of colors. GFP and its variants are frequently being used in creating sensors these days utilizing fluorescence resonance energy transfer (FRET) phenomenon. A receptor domain is fused with a donor and an acceptor fluorophore at N and C terminus, respectively. On sensing a ligand, the receptor undergoes a conformational change resulting in FRET. More and more efforts are being made to utilize fluorescence technology in solving the underlying causes of a particular medical condition. These genetically encoded FRET-based sensors can be helpful in the real-time analysis of the cells permitting visualization of complex intracellular protein pathways, genetic and cytoskeletal dynamics. Many diseased conditions arise because of the disruption of a particular pathway, changes in the concentration level of a particular metabolite, or under- or overexpression of a specific gene. These FRET-based nanosensors help in visualizing various cellular and subcellular events in vivo that might be playing a role in certain pathological disorders. These sensors have not only widened the field of medical research but also has created a hope for early detection and prevention of a disease or a disorder.

N. Soleja · M. Mohsin (✉)
Department of Biosciences, Jamia Millia Islamia, New Delhi, India
e-mail: mmohsin1@jmi.ac.in

M. Mohsin et al. (eds.), *Nanobiosensors for Agricultural, Medical and Environmental Applications*, https://doi.org/10.1007/978-981-15-8346-9_1

Keywords

Fluorescent proteins · FRET · Genetically encoded · Nanosensor · Periplasmic-binding proteins

Abbreviations

FP	Fluorescent proteins
FRET	Fluorescence resonance energy transfer
GFP	Green fluorescent protein
PBP	Periplasmic-binding protein
QDs	Quantum dots

1.1 Introduction

Various neurological disorders (multiple sclerosis, Parkinson's disease, Alzheimer's disease), diabetes, heavy metal toxicity, cardiovascular diseases, cancer, major vitamin deficiencies, muscular dystrophy, inflammatory diseases, skeletal disorders pose a major threat to humankind. Early detection, diagnosis, and timely treatment of the diseases can be beneficial. Despite significant progress in the regulatory networks that control metabolism, we have very little knowledge about the dynamic fluctuations of metabolite levels in time and space. With the rapid growth in the drug discovery programs, it has become essential to visualize and monitor biological activities in healthy and diseased cells, with high spatial and temporal resolution (Morris 2010; Deuschle et al. 2005).

The development of optical probes and sensors which can report on the cellular concentrations and activities of specific intracellular targets has become crucial. Nuclear magnetic resonance (NMR), atomic absorption spectroscopy, atomic force microscopy, mass spectroscopy, and various other techniques are being employed to detect molecular events and interactions in a cell, but most of them could not monitor and visualize dynamic changes in real time (Mohsin et al. 2015). Methodologic limitations and analytical problems of reproducibility and standardization make these approaches quite complicated, which is also labor-intensive and requires tissue fractionation and sample preparation (Jenčo et al. 2017). A substantially higher resolution has been achieved with the recent developments in the field of fluorescent biosensor technology. Fluorescent probes have the advantage of providing subcellular resolution, accurate sensitivity, high selectivity, and versatility (Deuschle et al. 2005). Earlier, molecular organic dyes and quantum dots (QDs) were exploited as optical probes, but due to their invasive nature, the focus was shifted to fluorescent genetic reporters (Okumoto 2010). Together with the discovery of GFP, development of fluorescence-based reporters, and advent of novel imaging techniques, the biosensor field has made possible high throughput screening and analysis of complex biological samples, cellular and molecular interactions. Moreover, the use of two suitable fluorophores that can undergo FRET in the development of genetically encoded nanosensors permits ratiometric and thus the quantitative analysis of the metabolites in living cells (Morris 2010; Soleja et al. 2018). These sensors exploit the physiological changes in a cell to convert it into an optical signal in the form of FRET ratio which is defined as the change in the emission ratio of an acceptor to the donor. Visualizing cellular and subcellular distribution of molecules within a cell, their transport, flux, interactions with the environment, molecular crowing, structural, and functional information is possible with the help of these sensors (Tsien 1998; Lippincott-Schwartz and Patterson 2003).

In this article, recent advances in the development of genetically encoded nanosensors are given and their application in the understanding of metabolic and regulatory networks to elucidate the mechanisms responsible for molecular fluxomics have been discussed.

1.2 Fluorescent Proteins and FRET Phenomenon

Fluorescent proteins (FPs) display bright fluorescence in the entire visible spectrum and have simplified the task of live imaging cells, deep tissue, and whole-body in vivo with minimal perturbation in a non-destructive manner (Zhang et al. 2002; Pakhomov and Martynov 2008; Okumoto 2010). GFP is one such fluorescent protein whose remarkable discovery led the scientists to think of ways to "paint" a wide array of biological molecules with various colors at the same time (Tsien 1998). With plenty of mutants, GFPs are now important tools in current cell biology research covering blue to the yellow range of the electromagnetic spectrum and has facilitated real-time visualization of metabolite levels.

FPs are extensively being used in bio-imaging studies due to the following characteristics: (a) They do not exhibit cellular toxicity and can be expressed very efficiently within the living cells, (b) FPs do not oligomerize when co-expressed, and (c) they are photostable for an extended period at physiological parameters (Soleja et al. 2018).

FRET is a spectroscopic process by which energy is transferred in a non-radiative manner between donor and acceptor molecules through long-range dipole–dipole coupling interaction. This energy transfer leads to a reduction in the donor's fluorescence intensity and an increase in the acceptor's emission intensity. FRET efficiency depends on (a) the distance between the two fluorophores, (b) spectral overlap between the donor's emission and acceptor's absorbance, (c) parallel orientation of the two dipoles, (d) high-quantum yield of donor, and (e) high-absorption coefficient of acceptor (Zhang et al. 2002; Jares-Erijman and Jovin 2003) (Fig. 1.1). FRET is primarily being used to determine the separation distance between the two FPs or to measure a structural change as a result of biological perturbation.

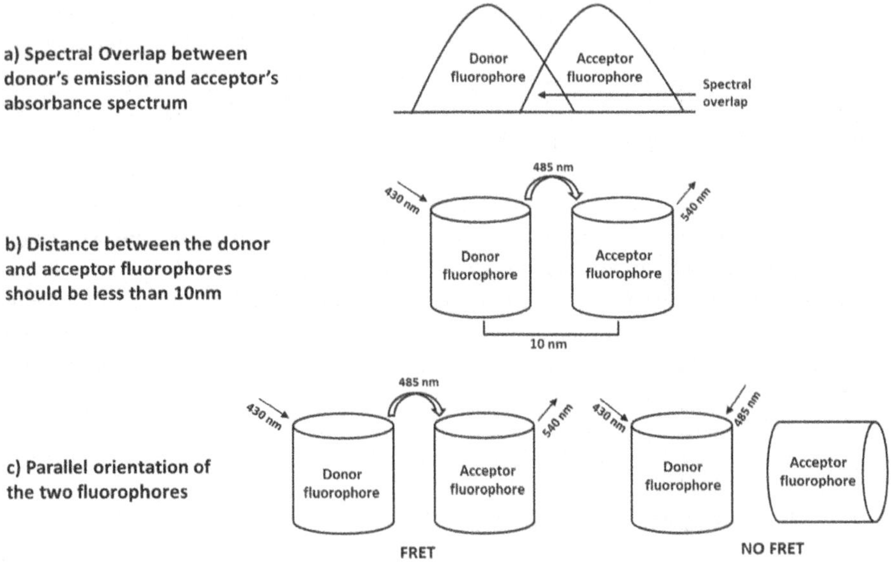

Fig. 1.1 Conditions of FRET

1.3 Strategies for Developing Genetically Encoded FRET-Based Nanosensors

The discovery of fluorescent proteins, along with the advent of bio-imaging technology, has revolutionized cellular and molecular biology, providing insight into the clinical and medical research. Usually, three strategies are adopted for designing a FRET-based nanosensor (Palmer et al. 2011) (Fig. 1.2).

(a) Metabolite binding to a target molecule: For the construction of a FRET-based nanosensor, the ligand-binding domain is sandwiched between donor and acceptor fluorophores. Bacterial periplasmic-binding proteins (PBPs) are suitable candidates for ligand sensing domains and recognize a number of substrates with high affinity and specificity. Upon ligand binding, PBPs undergo significant conformational changes that increase the FRET efficiency between the two FPs. Cameleon Ca^{2+}-sensors, as well as sensors for sugars, glutamate, leucine, cAMP, cGMP, NO, and membrane potential fall under this category.
(b) Enzyme-dependent (protease-activated) FRET-based sensors: Proteins, when activated by enzyme, cleaves a certain sequence that leads to a reduction of FRET between the donor and acceptor as the distance of FPs increases from each other like caspases and matrix metalloproteases.
(c) Protein–protein interaction: In such types of sensors, the receptor domain is tagged with a donor fluorophore and hormone with an acceptor fluorophore. When the hormone binds to the receptor, the two fluorophores come close together, resulting in FRET.

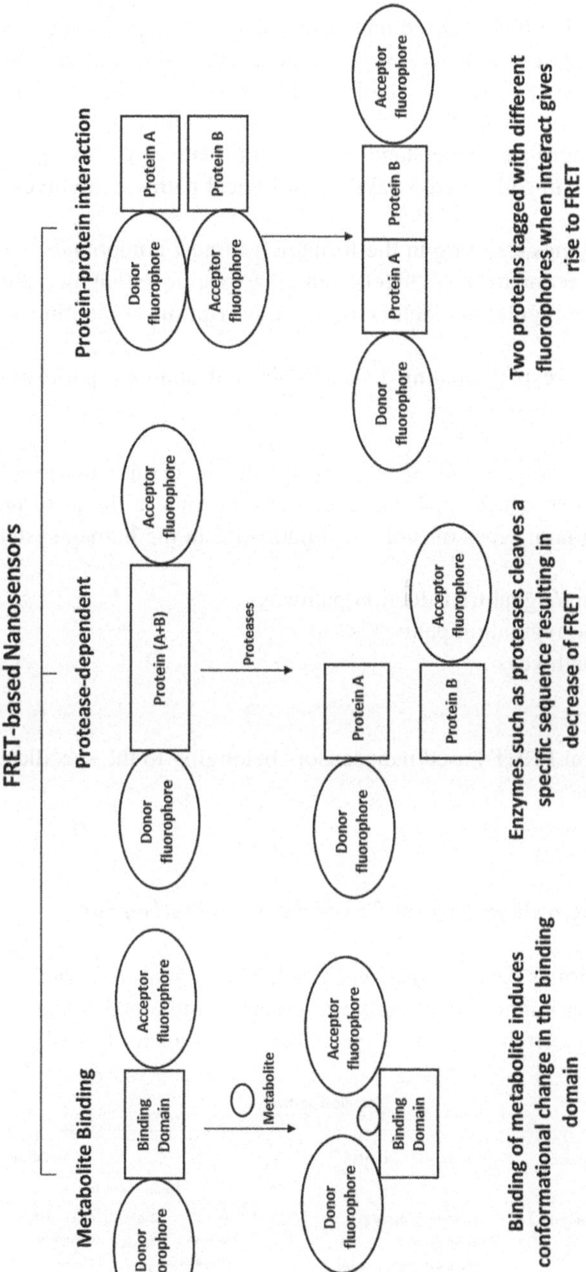

Fig. 1.2 Strategies for designing FRET-based nanosensors

1.4 Role of FRET-Based Nanosensors in Biological Pathways

Genetically encoded FRET-based nanosensors constitute promising and convenient means of probing and reporting on the presence, level, and activity of specific metabolites within the complex biological environment in real time in vitro, *in cellulo*, and in vivo with high spatio-temporal resolution (Morris 2010). Such sensors help explore and understand biological processes involving complex intracellular pathways in cell-based assays. A biological pathway involves

(a) Certain enzymes resulting in the formation of new compounds
(b) Primary or secondary messengers inducing specific signaling pathways
(c) Certain gene-regulating factors (e.g., transcription) switching on and off a gene, and
(d) Metabolites are participating in catabolic and anabolic pathways controlling energy flow.

Any sort of disruption in a pathway results in certain diseases, syndromes, disorders, and even cancer that can eventually become a threat to human health. There are three major types of biological pathways in the human system (Fig. 1.3).

(a) Cell signaling/signal transduction pathways
(b) Genetic/gene regulation pathways
(c) Metabolic pathways

Some common FRET-based nanosensors belonging to the specific pathways are discussed herein.

1.4.1 Cell Signaling/Signal Transduction Pathways

Signal transduction is the transmission of molecular signals (exterior or interior) into a cell, which then initiates an effective response. It initiates a cascade of molecular events involving certain hormones, secondary messengers (cAMP, cGMP, DAG,

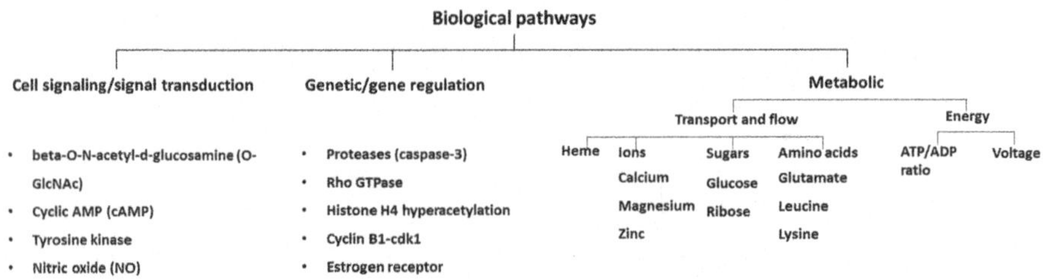

Fig. 1.3 Types of biological pathways

GTPases), enzymes and its cofactors, pro-apoptotic, and anti-apoptotic signals. Genetically encoded FRET-based nanosensors help to provide an insight into the various molecular events involved in cell signaling. Some genetically encoded FRET-based sensors for cell signaling pathways are as follows.

1.4.1.1 Beta-O-N-Acetyl-d-Glucosamine (O-GlcNAc)

O-GlcNAc is known to modify several proteins in the cell. It plays a role in the regulation of various autoimmune disorders, cardiac and vascular dysfunction, and hypertension. Signals such as high glucose level and cellular damage dynamically change O-GlcNAc levels in the cell, causing Alzheimer's and type II diabetes. Genetically encoded FRET-based nanosensor was developed by sandwiching a binding domain, GafD fused to a substrate within the enhanced cyan fluorescent protein (eCFP) and Venus. Binding of GlcNAc brings the two fluorophores closer, resulting in FRET (Carrillo et al. 2006).

1.4.1.2 Cyclic AMP (cAMP)

Binding of molecules such as hormones, prostaglandins to G protein-coupled receptors (GPCR) results in adenylyl cyclase activation which in turn produces cyclic adenosine monophosphate (cAMP). It is known to regulate various inflammatory responses. Disruption of cAMP related pathways is linked with certain cancers and might play some role in diabetes. It suggested that cAMP relieves the interaction between the regulatory and catalytic domains of Epac. Epac1 is a guanine nucleotide exchange factor for Rap1 that is activated by the direct binding of cAMP. They monitored Epac1 activation in vivo by using a cyan fluorescent protein (CFP)–Epac–yellow fluorescent protein (YFP) fusion construct. FRET rapidly decreases in response to the cAMP-raising agents, whereas it fully recovers after the addition of cAMP-lowering agonists (Ponsioen et al. 2004).

1.4.1.3 Tyrosine Kinase

Tyrosine kinases, as the name suggests, phosphorylate serine and tyrosine residues of certain growth factors. They are known to activate lymphocytes involved in immunological responses. This enzyme also has some cancerous effects related to the growth of malignant tumors. Three genetically encoded fluorescent reporters were constructed for the tyrosine kinases Src, Abl, and epidermal growth factor (EGF) receptor. The reporters had CFP, a phosphotyrosine binding domain, a consensus substrate for the relevant kinase, and YFP. The addition of growth factors stimulated the kinase activities in the living cell showing a 20–35% increase in the ratios of yellow to cyan emissions because phosphorylation induced changes resulted in FRET (Ting et al. 2001).

1.4.1.4 Nitric Oxide (NO)

NO is a neurotransmitter and a vasodilator released by macrophages in response to infections. A sensor was generated consisting of metallothionein (MT) sandwiched between two mutant GFPs to directly study the interaction between MT and NO in live cells. This construct (FRET-MT) was used to detect FRET following conformational changes in the periplasmic-binding protein, indicating the metal release from MT (Pearce et al. 2000).

1.4.2 Genetic/Gene Regulation Pathways

Gene regulation involves certain factors that control the level of gene expression at the RNA and protein levels. The follow-up of diseases due to genetic alterations, gene rearrangement, less basal level of expression is possible because of these FRET-based sensors.

1.4.2.1 Proteases (Caspase-3)

Caspases are linked with the development of tumors. Mutation in cell cycle genes and mutated or ineffective caspases can result in tumor formation or autoimmune diseases. Overexpression of caspases can also be a problem resulting in excessive programmed cell death (e.g., neurodegenerative diseases such as Alzheimer's disease). A genetically encoded construct was prepared by linking a CPP32 recognition sequence, DEVD to GFP, and blue fluorescent protein (BFP), respectively, at its N and C termini. During apoptosis, this protease cleaves the recognition sequences, which results in the loss of FRET efficiency (Xu et al. 1998).

1.4.2.2 Rho GTPase

Rho family of GTPases is known to be participating in certain circulatory, lung, neurological, and cytoskeletal disorders and even cancer. Rho GTPases are a subfamily of the Ras proteins controlling cell's dynamics: morphology, division, function. A genetically encoded FRET-based nanosensor was developed for Cdc42 and Cdc42 guanine nucleotide exchange factors (GEFs) by ligating CFP and YFP to a Rho protein, Cdc42 (induces a conformational change in its downstream effector, the Wiskott–Aldrich syndrome protein (WASP)) at N and C termini, respectively. 3.2-fold changes in the FRET ratio were observed in the case of Cdc42, whereas GEF sensor upon exchanging of GDP for GTP shows a 1.7-fold change (Yoshizaki et al. 2003).

1.4.2.3 Histone H_4 Hyperacetylation

Acetylation of histone H_4 helps in maintaining the nucleosome structure. Defect in nucleosome structure is reported to cause various chromosomal abnormalities. To visualize histone acetylation in living cells, a five-part tandem fusion protein consisting of an acetylation-binding domain, a flexible linker, a substrate histone H_4, and the two different-colored mutants of GFP (CFP and Venus) was created for FRET-based analysis. It has now become feasible to monitor the dynamic fluctuation of histone H_4 acetylation levels during mitosis (Sasaki et al. 2009).

1.4.2.4 Cyclin B1-cdk1

During prophase, cyclin B1 gets imported into the nucleus, promoting nuclear envelope breakdown (NEBD) and disassembly of nucleoproteins and nuclear lamina. One of the DNA repair mechanism checks mitosis by inhibiting activation and nuclear import of cyclin B1-Cdk1. p21 acts as a cyclin-dependent kinase inhibitor (cdk inhibitor) to promote cell cycle arrest at G1/S transition state suppression through the inhibition of phosphorylation by cyclin-dependent kinases (cdks). p21 makes hematopoietic cells resistant to infection by aborting chromosomal integration of the HIV provirus. A FRET-based nanosensor was constructed using CFP and YFP variants to determine the level of cyclin B1-Cdk1 in living cells (Gavet and Pines 2010).

1.4.2.5 Estrogen Receptor

Estrogen is a female hormone responsible for the development of secondary sexual characteristics and regulating the menstrual and reproductive cycle in women. CFP and YFP variants were attached to the ER ligand-binding domain (LBD), respectively, at the N and C termini to detect estrogen receptor (ER) ligands. The wild-type LBD on binding with the ligand undergoes a conformational change to cause FRET. Ala 430 was mutated to Asp to cause a four-fold increase in the FRET signal. Truncated YFP and eCFP were used to determine the affinity of this sensor. Also, FRET can be applied to various nuclear receptors, such as the androgen receptor (De et al. 2005).

1.4.3 Metabolic Pathways

Ions, molecules, amino acids, sugars, energy flow involved in various metabolic processes can be monitored using these types of FRET-based sensors. Visualizing their compartmentalization, transport, flux, concentration in real time with high resolution is possible using the sensors.

1.4.3.1 Transport and Flow

Heme

Heme is thought to be an important participant in cellular processes like transcriptional regulation, protein localization/degradation, and iron homeostasis. Nonprotein-bound free heme is highly toxic to cells due to its inherent peroxidase activity. Two bacterial heme transfer chaperones are sandwiched by eCFP and eYFP at the N and C terminus. By systematically varying the linker length between the two heme chaperones, they obtained a heme probe with large ratiometric change upon heme binding to monitor intracellular heme and its regulation (Song et al. 2015).

Ions

Some ions act as important cofactors, while the excess of some may cause toxicity. There are many ion specific channels present in the cell responsible for regulating their balance in the cell.

Calcium Binding of Ca^{2+} to calmodulin makes it wrap around the M13 domain increasing the FRET signal between the flanking GFPs. Point mutations in calmodulin can increase the Ca^{2+} affinities for free Ca^{2+} ions in the physiological range 10^{-8}–10^{-2} M. It provides an insight into the regulatory roles played by this ion in real time, its transport and localization (Evanko and Haydon 2005).

Magnesium Limited access to the structural, catalytic, and signaling roles of magnesium led to the development of MagFRET-1, a sensor consisting of human centrin 3 (HsCen3) fused with cerulean and citrine fluorescent domains at the N and C terminus with 50% increase in emission ratio and binds magnesium with $K_d = 148$ mM. It helped in understanding intracellular Mg^{2+} homeostasis and signaling (Lindenburg et al. 2013).

Zinc A FRET-based nanosensor was developed using the cyanobacterial metallothionein SmtA flanked by CFP and YFP to analyze the uptake and metabolism of zinc in *Escherichia coli (E. coli)*. Zinc is essential for DNA synthesis, transcription, and translation, but an excess of it can lead to toxicity. On binding of the zinc with the periplasmic-binding protein in the construct, there was an increased FRET ratio with increasing Zn^{2+} concentration. Affinity mutants were constructed with binding constants ranging from 50 μM to 1 mM. In vivo analysis of this nanosensor in *E. coli* showed that free Zn^{2+} accumulates in the bacterial cells and is slowly metabolized (Mohsin et al. 2015).

Sugars

Sugars provide energy for cellular processes.

Glucose To monitor glucose homeostasis, transport across the plasma membrane and its metabolism, a glucose nanosensor (FLIPglu) was developed by flanking the *E. coli* periplasmic glucose/galactose-binding protein with two different green fluorescent protein variants. Substrate binding showed a concentration-dependent decrease in FRET between the attached GFP variants with a binding affinity for glucose of 170 nM. An affinity mutant with a $K_d = 600$ μM was generated, which showed higher substrate specificity (Fehr et al. 2003).

Ribose This sugar is essential for the formation of both DNA and RNA. A ribose biosensor was first constructed by covalently attaching fluorescent probes to RBP. It displayed a 54% decrease in emission intensity upon the addition of saturating ribose concentrations and exhibited a dissociation constant $K_d = 3.4$ mM. Site-directed mutants were created for examining ribose binding ability and structural stability (Lager et al. 2003).

Amino Acids

Amino acids are the building blocks of protein that plays a crucial role in almost all biological processes. Also, they constitute a large proportion of muscles and tissues.

Glutamate Alzheimer's disease, epilepsy, and hepatic disorders are associated with this amino acid. FLIPE was constructed by sandwiching the binding domain ybeJ between two GFP variants. FLIPE shows a concentration-dependent decrease in FRET efficiency in the presence of ligands. A decrease in FRET ratio was observed during depolarization, whereas no change was observed when this sensor was expressed in the cytosol (Okumoto et al. 2005).

Leucine Considering the role of this amino acid in synthesis and translational control of muscle proteins, energy metabolism, and metabolic disorders associated with its loss such as type II diabetes and muscular dystrophy, a genetically encoded FRET-based sensor was constructed by combining LivK (a PBP) with CFP and YFP as donor and acceptor fluorophores, respectively. It showed a concentration-dependent change in FRET ratio covering the physiological range between 8 μM and 01 mM (Mohsin et al. 2013).

Lysine Lysine is an antagonist of serotonin reducing signs of anxiety. Deficiency of lysine causes an increase in serotonin concentration regulating emotions and stress. Lysine binding periplasmic protein (LAO) from *Salmonella enterica* serovar Typhimurium was used as a binding protein fused with CFP and YFP at N and C termini. The sensor was named as FLIPK with an affinity constant $K_d = 97$ μM (Ameen et al. 2016).

1.4.3.2 Energy

ATP/ADP Ratio

Most of the metabolic functions such as synthesis of proteins, membranes, DNA, RNA intracellular, and extracellular signaling require energy. A ratiometric fluorescent sensor was created by fusing the bacterial regulatory protein GlnK1 from *Methanococcus jannaschii* to a circularly permuted fluorescent protein to monitor the ATP/ADP concentration ratio in living cells. The affinity for Mg-ATP was found to be 100 nM. ADP binds to the same site, competing with Mg-ATP but producing a smaller change in fluorescence (Berg et al. 2009).

Voltage

Maintenance of potential across the membranes is necessary to control the flux of various ions and metabolites in the cell, such as sodium and potassium channels. Skeletal and neurological syndromes are associated with the disruption in the membrane potential. The sensor protein, VSFP1, is made up of sensor domain from the Kv2.1 potassium channel ligated with CFP and YFP as donor and acceptor fluorophores. It shows a concentration-dependent change in FRET with membrane potential (Dimitrov et al. 2007).

1.5 Limitations of Fluorescent Proteins and Possible FRET Challenges

(a) The fusion of FPs as molecular tags increases the overall size of the protein that offers steric hindrances for protein folding and their functions and expressions.
(b) Improper choice of FPs may result in "unplanned" homo-fretting, spectral crosstalk/bleed-through, and background autofluorescence problems.
(c) The rate of chromophore maturation of different FPs (GFP shows slow maturation) may provide constraints on the utility of these molecules.
(d) FPs can aggregate leading to cellular toxicity. Also, the excitation of GFP for a longer period results in the generation of free radicals that are toxic to the cells.
(e) The detection of low expression levels becomes difficult because of the uncontrolled amplification of the GFP signal.
(f) FPs like YFP are sensitive to halide ion concentration, while some show pH sensitivity and are thermally unstable (Jensen 2012; Soleja et al. 2018).

1.6 Conclusions and Future Perspectives

In recent years, metabolic engineering has emerged as an effective tool for studying, enhancing, and modifying multiple cellular and metabolic pathways using recombinant DNA technology and innovative experimental techniques focused on determining concentrations, metabolic fluxes, and their in vivo regulatory control. The challenge is understanding the precise cellular and subcellular metabolic networks in a highly compartmentalized multi-cellular organism. To better understand the underlying causes of severe pathologies, it is important to monitor the dynamic changes in the distribution of metabolites at a single-cell level. A wide range of biophysical techniques have been used in the past to quantify metabolomic content in a cell, covering the genetic and biochemical dimensions of metabolic flux modeling, but they suffer from various technical and methodological constraints. These classical methods are somewhat intrusive, costly, bio-incompatible, complex, time-consuming, static, provide weak dynamic data, and are susceptible to objects that further reinforce the need to discontinue their use for single-cell analysis. Advanced techniques are needed to give a better understanding of regulatory pathways/processes, metabolite transport, compartmentalization, and sensing.

Fluorescence-based biosensors are promising, rapidly emerging, and cost-effective tools for system-wide characterization and have achieved tremendous recognition among scientists over the past decade. This sensor-backed technology works by identifying and tracking metabolite distribution within the cellular compartments and in real time assessing the precise concentration of these metabolites at trace level. Genetically encoded FRET-based nanosensors use fluorescent proteins as reporters and calculate the dynamics of single-cell concentrations of biophysical processes, ions, and metabolites by leveraging the necessary conformational changes of periplasmic-binding proteins. To resolve biologically relevant issues such FRET-based sensors are needed. These sensors help to understand the derivative link behind the underlying metabolite imbalance mechanisms and an individual's resulting toxicological profile. Compared to earlier techniques, these nanosensors provide real-time monitoring and measurement of biological molecules at intervals of milliseconds for many hours.

Acknowledgments The first author (NS) gratefully acknowledges the UGC, Govt. of India, for providing the financial support in the form of Junior Research Fellowship.

References

Ameen S, Ahmad M, Mohsin M, Qureshi MI, Ibrahim MM, Abdin MZ, Ahmad A (2016) Designing, construction and characterization of genetically encoded FRET-based nanosensor for real time monitoring of lysine flux in living cells. J Nanobiotechnol 14:49

Berg J, Hung YP, Yellen G (2009) A genetically encoded fluorescent reporter of ATP:ADP ratio. Nat Methods 6:161–166

Carrillo LD, Krishnamoorthy L, Mahal LK (2006) A cellular FRET-based sensor for beta-O-GlcNAc, a dynamic carbohydrate modification involved in signalling. J Am Chem Soc 128:14768–14769

De S, Macara IG, Lannigan DA (2005) Novel biosensors for the detection of estrogen receptor ligands. J Steroid Biochem Mol Biol 96:235–244

Deuschle K, Fehr M, Hilpert M, Lager I, Lalonde S, Looger LL, Okumoto S, Persson J, Schmidt A, Frommer WB (2005) Genetically encoded sensors for metabolites. Cytometry A 64:3–9

Dimitrov D, He Y, Mutoh H, Baker BJ, Cohen L, Akemann W, Knöpfel T (2007) Engineering and characterization of an enhanced fluorescent protein voltage sensor. PLoS One 2:e440

Evanko DS, Haydon PG (2005) Elimination of environmental sensitivity in a cameleon FRET-based calcium sensor via replacement of the acceptor with Venus. Cell Calcium 37:341–348

Fehr M, Lalonde S, Lager I, Wolff MW, Frommer WB (2003) In vivo imaging of the dynamics of glucose uptake in the cytosol of COS-7 cells by fluorescent nanosensors. J Biol Chem 278:19127–19133

Gavet O, Pines J (2010) Progressive activation of CyclinB1-Cdk1 coordinates entry to mitosis. Dev Cell 18:533–543

Jares-Erijman EA, Jovin TM (2003) FRET imaging. Nat Biotechnol 21:1387–1395

Jenčo J, Krčmová LK, Solichová D, Solich P (2017) Recent trends in determination of thiamine and its derivatives in clinical practice. J Chromatogr A 1510:1–12

Jensen EC (2012) Use of fluorescent probes: their effect on cell biology and limitations. Anat Rec (Hoboken) 295:2031–2036

Lager I, Fehr M, Frommer WB, Lalonde S (2003) Development of a fluorescent nanosensor for ribose. FEBS Lett 553:85–89

Lindenburg LH, Vinkenborg JL, Oortwijn J, Aper SJA, Merkx M (2013) MagFRET: the first genetically encoded fluorescent Mg^{2+} sensor. PLoS One 8:e82009

Lippincott-Schwartz J, Patterson GH (2003) Development and use of fluorescent protein markers in living cells. Science 300:87–91

Mohsin M, Abdin MZ, Nischal L, Kardam H, Ahmad A (2013) Genetically encoded FRET-based nanosensor for *in vivo* measurement of leucine. Biosens Bioelectron 50:72–77

Mohsin M, Diwan H, Khan I, Ahmad A (2015) Genetically encoded FRET-based nanosensor for *in vivo* monitoring of zinc concentration in physiological environment of living cell. Biochem Eng J 102:62–68

Morris MC (2010) Fluorescent biosensors of intracellular targets from genetically encoded reporters to modular polypeptide probes. Cell Biochem Biophys 56:19–37

Okumoto S (2010) Imaging approach for monitoring cellular metabolites and ions using genetically encoded biosensors. Curr Opin Biotech 21:45–54

Okumoto S, Looger LL, Micheva KD, Reimer RJ, Smith SJ, Frommer WB (2005) Detection of glutamate release from neurons by genetically encoded surface-displayed FRET nanosensors. Proc Natl Acad Sci U S A 102:8740–8745

Pakhomov AA, Martynov VI (2008) GFP family: structural insights into spectral tuning. Chem Biol 15:755–764

Palmer AE, Qin Y, Genevieve J, McCombs JE (2011) Design and application of genetically encoded biosensors. Trends Biotechnol 29:144–152

Pearce LL, Gandley RE, Han W, Wasserloos K, Stitt M, Kanai AJ, McLaughlin MK, Pitt BR, Levitan ES (2000) Role of metallothionein in nitric oxide signaling as revealed by a green fluorescent fusion protein. Proc Natl Acad Sci U S A 97:477–482

Ponsioen B, Zhao J, Riedl J, Zwartkruis F, van der Krogt G, Zaccolo M, Moolenaar WH, Bos JL, Jalink K (2004) Detecting cAMP-induced Epac activation by fluorescence resonance energy transfer: Epac as a novel cAMP indicator. EMBO Rep 5:1176–1180

Sasaki K, Ito T, Nishino N, Khochbin S, Yoshida M (2009) Real-time imaging of histone H4 hyperacetylation in living cells. Proc Natl Acad Sci U S A 106:16257–16262

Soleja N, Manzoor O, Khan I, Ahmad A, Mohsin M (2018) Role of green fluorescent proteins and their variants in development of FRET-based sensors. J Biosci 43:763–784

Song Y, Yang M, Wegner SV, Zhao J, Zhu R, Wu Y, He C, Chen PR (2015) A genetically encoded FRET sensor for intracellular Heme. ACS Chem Biol 10:1610–1615

Ting AY, Kain KH, Klemke RL, Tsien RY (2001) Genetically encoded fluorescent reporters of protein tyrosine kinase activities in living cells. Proc Natl Acad Sci U S A 98:15003–15008

Tsien RY (1998) The green fluorescent protein. Annu Rev Biochem 67:509–544

Xu X, Gerard AL, Huang BC, Anderson DC, Payan DG, Luo Y (1998) Detection of programmed cell death using fluorescence energy transfer. Nucleic Acids Res 26:2034–2035

Yoshizaki H, Ohba Y, Kurokawa K, Itoh RE, Nakamura T, Mochizuki N, Nagashima K, Matsuda M (2003) Activity of Rho-family GTPases during cell division as visualized with FRET-based probes. J Cell Biol 162:223–232

Zhang J, Campbell RE, Ting AY, Tsien RY (2002) Creating new fluorescent probes for cell biology. Nat Rev Mol Cell Biol 3:906–918

Current Status of Nanosensors in Biological Sciences

2

Ghazala Ambrin, Habiba Kausar, Ruphi Naz, and Altaf Ahmad

Abstract

Nanotechnology is an emerging tool in the field of life sciences and has since continually proven its potential in contributing to sustainable competitiveness and growth in several industrial sectors. Though we have only managed to scrape the surface of the potential impact of nanotechnology in the coming years, the noteworthy findings, however, cannot be neglected. Recent years have seen a revolutionizing improvement in industrial sectors which has stemmed from the development of various nanosensors in an overwhelming proportion employing their ability to detect chemical or biological species on a nanoscale. A detailed study of nanosensors and its various types, namely, magnetic, chemical, mechanical, optical, thermal, and nanobiosensors, and their fabrication and application in the field of biotechnology has been discussed in the chapter allowing the reader to gain a better understanding of the new striking advancements in the field of research.

Keywords

Nanosensor · FRET · Nanotechnology · Magnetic · Chemical · Optical · Mechanical · Thermal

G. Ambrin · H. Kausar · A. Ahmad (✉)
Department of Botany, Aligarh Muslim University, Aligarh, Uttar Pradesh, India
e-mail: aahmad.bo@amu.ac.in

R. Naz
Department of Biosciences, Jamia Millia Islamia, New Delhi, Delhi, India

M. Mohsin et al. (eds.), *Nanobiosensors for Agricultural, Medical and Environmental Applications*, https://doi.org/10.1007/978-981-15-8346-9_2

2.1 Introduction

Nanotechnology, the ever-growing, far-reaching area of research has not let any sphere of life untouched by its miraculous progress. One significant result of the advancements in nanotechnology is the development of nanosensors. These sensors are devices that exploit nanoscale interactions for the measure or the detection of the analyte in question. It has proven its potential in almost every sphere of science. The application of nanosensors has a tremendously beneficial impact on the physical, chemical, and biological stature of research. It has dramatically improved environmental quality, crop productivity, human health, pharmaceuticals, and medical technology. Besides its vast range of application in the biological sciences, nanosensors have a handful of utility in physical sciences as well. The chapter allows the reader to gain deep insights into the current fabrications of different nanosensors and its range of applications in biological sciences. Various nanosensors like magnetic, chemical, mechanical, optical, thermal, and nanobiosensors have been intensely discussed in the chapter providing readers a resource for a better understanding of nanotechnology and its application in basic life sciences.

2.2 Types of Nanosensors

Different types of nanosensors, namely, magnetic, nanobiosensor, mechanical, optical, and thermal sensors, have been discussed in the following sections.

2.2.1 Magnetic Nanosensors

Magnetic nanosensors are composed of magnetic nanoparticles used for biosensing due to their strong magnetic properties. It finds various applications in different fields of research, and the fabrication varies with the nature of the demand. Three basic types of magnetic nanosensors are discussed below.

2.2.1.1 Magnetic Relaxation Switch Assay Sensors

These nanosensors comprise magnetic nanoparticles or NPs (diameter 5–300 nm) or micrometer-sized magnetic particles or MPs (diameter 300–5000 nm) which are capable of detecting molecular interactions through magnetic resonance techniques (Koh and Josephson 2009). The spin–spin relaxation time or T_2 in nuclear magnetic resonance (NMR) study is defined as the time taken for the transverse magnetism to fall to approximately 37% of the initial value or by a factor of e, i.e., 2.718281828. When the magnetic nanoparticles encounter their specific molecular object, they lead to the establishment of stable nano-assemblies with the sizeable cross-sectional area, which further decreases the spin–spin relaxation time (T2) of the neighboring molecules (Josephson et al. 2001). MRSWs show variation in measured T2 values wherein Type I MRSw shows increased T2 value and Type II MRSw is related to a decreased T2 (Koh et al. 2009(a)). MRSWs make use of radiofrequency radiation for penetrating biological specimens irrespective of their optical properties (Perez et al.

2002). Therefore, experimentation with magnetic nanoparticles has been efficiently conducted in a lipid emulsion, turbid cell suspension, blood, tissue, and culture media as light plays no part in creating hindrances on the outcome of the assay (Perez et al. 2004). MPs or NPs aggregates are influenced by temperature, pH, and high analyte concentration resulting in the reversal of aggregation and hence called "relaxation switches." The spin–spin relaxation time or T2 value for aggregated and dispersed MPs or NPs varies among each other (Koh et al. 2009(b)). Magnetic resonance imaging uses T2 as a biological parameter for making a distinction between tissue types as the T2 readings for muscle, blood, water, and fat are 40 ms, 180 ms, 2500 ms, and 90 ms, respectively (Josephson et al. 2001). Several experiments have been conducted which exploit the technique of T2 measurement for detecting molecular targets like oligonucleotide sequences (Perez et al. 2002), endonucleases, viral particles in serum, and enzymatic activity (proteases) (Perez et al. 2004). Recently, applied magnetic relaxation sensor for the study of zika virus-host cell receptor interaction has paved the way for further experiments to unravel the different host–pathogen interactions in nature (Shelby et al. 2017).

2.2.1.2 Magnetic Particle Relaxation-Based Sensors

These sensors employ the relaxation of the magnetic moments aligned as a response to the turn-off of an applied magnetic field within magnetic particles as a basis for its assay. Semiconducting quantum interference devices (SQUIDs) are applied for the measurement of the relaxation of magnetic moments (Grossman et al. 2004). The magnetic moments have two relaxation mechanisms:

1. Brownian relaxation
2. Neel relaxation

1. Brownian relaxation sensors: Brownian relaxation of NPs has been used as biosensing through the measurement of static and dynamic susceptibility by the application of alternate currents. It involves the physical rotation of the entire particle and is much faster than the Neel relaxation. Aggregated NPs in solutions are more likely to show slower Brownian relaxation as compared to single NPs due to their larger hydrodynamic size (Tu et al. 2013). The change in the relaxation has been sensed in serum(Hong et al. 2006) and buffer with the help of SQUID or an ac magnetosusceptometer (Hong et al. 2007).

2. Neel Relaxation Sensors: The relaxation moment is reportedly measured by SQUID. The Neel relaxation time falls appropriately in the detection range of SQUID. However, the limit for its detection is ranged to be 5 x 10^4NPs for substrate assay and 1.1 x 10^5 bacteria in a sample volume of 20 μL (Lee et al. 2002).

2.2.1.3 Magnetoresistive Sensors

The mechanism for sensing involves the measurement of altered electric current ensued due to the change in the magnetic fields of both magnetic particle and the sensor upon binding of the former onto the sensor surface. There exist two methods for the binding of magnetic particles: (1) direct labeling and (2) indirect labeling. In indirect labeling, magnetic particles unite with the surface through complementary

DNA sequence recognition or streptavidin–biotin interaction. However, indirect labeling follows the ELISA sandwiched method where the target protein for a particular antibody is allowed for binding after the immobilization of the antibody, and further biotinylated antibodies are added to the system which is finally tagged by streptavidin-coated magnetic particles (Koh and Josephson 2009). Micrometer-sized magnetic particles have been utilized in magnetoresistive biosensing as they have the advantage of observation under a light microscope and also permits detection of an exceedingly small number of particles owing to higher particle-based magnetic moments. Recently, NPs replaced the use of MPs as they are less susceptible to particle clustering under magnetic field. Magnetic tunnel junction (MTJ) or giant magnetoresistance (GMR) spin valve sensors have been successfully applied for the detection of MPs. They are composed of several layers of ferromagnetic materials, and the surface for the attachment of biomolecules is prepared by the deposition of Au or SiO_2 layer (Wang and Li 2008). Spin valve sensors comprising streptavidin-coated MPs are reported to be used in the protein marker detection at 27 pg/ml level of sensitivity (De Palma et al. 2007a, b). Cancer marker detection in 50% serum at sub-picomolar concentrations has been demonstrated with the help of 50 nm MACS magnetic nanoparticles (Osterfeld et al. 2008). MR has also shown its potential in the electronics industry. Extraordinary magnetoresistance or EMR is a three-order higher variant of MR and could be applied for storing greater densities of information than giant magnetoresistance (Solin 2001). It could also offer enormous potential in the position-sensing robot and speed and position industry (Solin 2004).

2.2.2 Chemical Nanosensor

Chemical nanosensor detects the presence, concentration, or quantity of an analyte and produces a measurable signal output via a transducer. It makes use of electronics and capacitive readout cantilevers for the analysis of the signal (Mousavi et al. 2018), Chemical sensors follow concepts of *sensitivity,* i.e., detection of the nanomolar concentration of analytes, and *selectivity,* i.e., recognition of the molecular structure and their reactivity for processing. Sensitivity and selectivity are affected by the phase, temporal, and dimensional aspects of the desired determination. The analyte could be present at any dimensional scales ranging from liters to picoliters in solid, liquid, or gas phases. It may also be present at surface layers from nanoscopic to the monomolecular range. Different sets of requirements for designing of sensors can begin from a need of multiple estimations of the analyte over extended periods (days, months) or in remote areas, such as in the ecological examination and individual monitoring. Likewise, it also requires an evaluation for the needed level of quantitative precision and accuracy (Grate et al. 1993). The types of sensors could be discussed under two headings:

1. Direct-rendition, selective chemical sensors

The activity of a direct-perusing, selective chemical sensor depends on the presence of a specific recognition event that leads to a change in a quantifiable parameter. Mostly, the response to the presence of an analyte in the sample depends upon some form of chemical reactivity of the analyte. The molecular selectivity aspect commonly includes an intricate decision of the sensing chemistry and related materials (Brown et al. 1992). Potential candidates for sensing material include organic monolayers, polymers, nanostructured and porous materials, metal semiconductors, and biomolecules. Chemical reactivity includes a vast range of chemical anomaly, inclusive of recognition of size/shape/dipolar properties of analytes by any bioreceptor sites, ceramic or other materials with patterned cavities, or structures capable of allowing molecular recognition. The recognition enables the selective and robust binding of the analyte to the sensor (Haswell 1992).

2. A sensitive not necessarily selective detection involving chromatographic or electrophoretic step

Here, analytes are allowed to pass under pressure or electric current, leading to their separation through environmental interactions. The analytes are finally detected as it sequentially moves through the detector. Miniaturization of the sensor system is the most needed strategy to be followed for analysis as it is associated with great benefits. Sensor systems associated with miniaturization include fiber-optic sensor system, piezoelectric effect-based mass, electrochemical, amperometric, gas-liquid chromatography, HPLC system, SAW devices, and capillary-zone electrophoresis separation system (Monig and Jorgenson 1991).

As a general rule, the sensor mechanism follows the sensing of an analyte by the sensitive layer and transducing the physical or chemical alterations occurring at the sensing part for analysis. Different techniques for carrying out transduction could include optical, thermal, or electrochemical approach. The application for various transduction approach affected by changed and modified fabrication presents a wide range of sensing platforms. The interface between the transducer and external environment comprises the sensitive element, which is considered to be the heart of the chemical sensor.

Different chemical nanosensors have been developed, and its application in various field of researches has been reported. Metal-oxide semiconductors (MOSs), a class of chemiresistive sensors, have been used for automotive emission monitoring, environmental monitoring, and food safety testing owing to their feasible production cost, high sensitivity, ease in application, and capability for diverse detection types of gases. MOSs is categorized into n-type and p-type sensors characterized by electron and hole as the major carrier, respectively. MOS works on the principle of equilibrium shift of the interface reactions related to the target analyte. An increase in conductivity of n-type semiconductors and a contrasting decrease in conductivity of p-type semiconductors are observed due to the reducing gases such as CO, H_2, and NH_3, while an opposite effect is seen for oxidizing gases. Sensing elements of some gas chemical nanosensors have also been reported to be made of copper and nickel oxides. The fabrication of chemical nanosensors varies according to the principle of physics and mechanism involved inclusive of the sensing element (material and microstructures) being used and the integration of both material and sensor (Zhang et al. 2017). Carbon nanotubes (CNTs) are another popular and highly potent candidate for the gas sensor in recent years. CNTs are composed of nanocarbon material and are long-structured, perforated with graphene roll arranged walls. They usually have metal characteristic or semiconductors associated with the tube's diameter and the rotational angle. Two forms of CNTs are known: (1) SWCNT (single-walled carbon nanotube), which is made from one roll of graphene layer, and (2) MWCNT, which comprises the concentric tube. The advantage of CNTs as the gas sensor can be ascribed to the large surface area, multiple absorption points, nanosize, and its high sensitivity to the surrounding chemical environment (Schroeder et al. 2019).

Several nanosensors measure the related physical parameters through the alteration in the resonant frequency of an oscillating piezoelectric crystal. Quartz crystal microbalance (QCM) or surface acoustic wave (SAW) sensors (Penza et al. 2004) are classified based on the kind of the vibrational wave circulated. Their sensitive layer is typically composed of a thin polymeric film; however, unconventional materials have also been reported in use. Chemical sensors based on the optical properties have progressed through the time of the first sensor developed for measuring O_2 and CO_2 concentrations. pH measurement through opto-chemical nanosensors has revolutionized the field of sensor research. pH measurement was made possible inside the extra-embryonic space of a rat conceptus through the insertion of nanosensors without damaging the visceral yolk sac (Modi et al. 2003). Moreover, chloride and nitrite levels in the yolk sac of rat conceptus were also established. Na^+ concentration in the cytoplasmic space of a mouse oocyte has been made possible through opto-chemical nanosensors. The concentration was measured during the period of opening and closing of ion channels under kainic acid used as an external stimulant. Likewise, Ca^{2+} concentration in single cells was also measured through the insertion of nanosensors into stimulated vascular smooth muscle cells. The measured Ca^{2+} fluctuations were interrelated to cellular stimulation (Cusano et al. 2008). A vast range of devices is developed on the chemical sensing mechanism wherein the analyte quantity is determined through the measurement of absorbance, refractive index, and fluorescence properties of the analyte. Such devices include spectroscopy (Raman, phosphorescence, luminescence, fluorescence), ellipsometry, surface plasmon resonance, and interferometry (Wang et al. 2017). Optical fibers are also a thorough source for transduction and have been exploited for the establishment of optical fiber sensors due to their profound property of transmitting light to more considerable distances with minimal loss helping in unraveling remote localization of the sensor head. The phenomenon works brilliantly for the hazardous or toxic chemicals present in the environment and also protects from the likelihood of gas explosion through otherwise electrical signals transduced by other chemical sensors. Fiber-optic chemical sensors offer a remarkable sensing platform since the wavelength, polarization, phase, and intensity of the light can all be used as parameters for measurement. Also, the different wavelengths released in either direction in the same fiber produce an autonomous signal. Therefore, different chemicals with the same fiber sensor could easily be monitored (Vallejo and Amo 2012).

2.2.3 Nanobiosensors

A biosensor is a tool for detection of target biological molecule with the help of bio-recognition elements that transduces the generated signal for analysis. It usually comprises a biological recognition element, a transducer, and the signal processor. A simple working of biosensors includes a sequential series of binding of the sample to the bioreceptors or the recognition element, initiation of the biochemical reactions at the electrochemical interface, conversion of the responses to electrical signals by transducers, translation of the electrical signal into expressive physical parameters, and display of the results to the operator (Jianrong et al. 2004). Nanobiosensors employ nanomaterials for the fabrication of the recognition element or transducers of biosensors. The nanomaterials include magnetic nanoparticles, nanotubes, quantum dots, metal nanoparticles, oxide nanoparticles, metallophthalocyanines, or other biological nanomaterials. The submicron size of these nanomaterials has improved and eased the quick biological and chemical analysis in vivo. In recent years, different properties of nanomaterials have been put to application, e.g., high speed, small size, lower power, less distance to travel by electrons, and low voltage (Kissinger 2005).

Nanobiosensor considers the sensitivity, specificity, and dynamic range of the sensor for enhanced functionality. The sensitivity is best described as the total output signal generated per analyte concentration or the minimum quantity of the analyte resolved by the sensor over a background signal better called a limit of determination (LOD). Another feature for sensitivity includes the accuracy for accurate analyte detection to differentiate between the true-positive and false-negative results (Cooper 2009). The dynamic range of nanobiosensors is a rarely discussed topic as it may or may not be of significant value depending on the application. However, it can be defined as the appropriate range of detection by the sensor to produce the accurate signal output. The specificity of a sensor validates the presence of the target analyte amidst other unknown materials. The feature plays an especially vital role in experimentation setups containing a low concentration of target analyte and high chances of nonspecific binding of the predominant unknown element to the sensor, helping to rule out the possibility of unusual signals (Bellan et al. 2011). The ligands bind to the surface of the sensor through specific interactions found in nature like nucleic acid hybridization, biotin-streptavidin, antibody–antigen, and enzyme–substrate interaction. Synthetic ligands like aptamers have also been applied in several cases, where nucleic acids ligands are artificially synthesized to bind to a wide range of host sensors through a process of systematic evolution of ligands by exponential enrichment (SELEX) (Ikebukuro et al. 2005). Molecularly imprinted polymers are another class of synthetic ligands produced to enhance the specificity to target sensor. Binding of the ligand to the sensor surface depends mainly on the surface chemistry. Physical absorption of target molecule mostly involves noncovalent binding. Gold nanoparticles utilize a free thiol group linked to a functional group for molecular recognition (Shankaran et al. 2007). Silanized silicon surfaces bind target molecules through different functional groups like amines as seen in APES (Howarter and Youngblood 2006), thiols, and epoxides. The magnetic nanoparticle usually has a coat of dextran or polyethylene glycol (PEG) that is capable of interacting with functional group of target molecules like aldehydes, amines, and carboxylic acids to form strong covalent bonds for immobilization of targets (Nitin et al. 2004). Another technique for probe–host interaction includes the use of CNT. Proteins and DNA have been reported to non-specifically bind to the CNT surface where CNT sidewalls interact with the aromatic molecules via π-stacking (Hirsch and Vostrowsky 2005).

The field of nanobiosensor encompasses and requires expertise from various backgrounds such as medical science, physics, chemistry, biological science, engineering, and so on for accomplishment in developing different novel biosensors (Mahato et al. 2016). A lot of effort has been put into the construction of distinct types of nanobiosensor forms as seen by the advanced built plastic antibodies (Chandra et al. 2015), synthetic ligands (Kashefi-Kheyrabadi et al. 2014), catalytic electrodes (Noh et al. 2012), nanotechnologies improvements for cell surface interaction (Blind and Blank 2015), and devices for medical applications (Turner 2013). Researchers have presented the marvel of interference-free endotoxin detection systems, paper inkjet sensor (da Silva et al. 2014), immunosensor (Mars et al. 2016), enzyme sensor (Kurbanoglu et al. 2015), microfluidic sensor devices (de la Escosura-Muñiz and Merkoçi 2016), etc. The detection of cancer, heavy metal ions, bacteria, hormones, and parasites has been dealt with the improvement in nanobiotechnology (Mahato et al. 2016). Reports suggest the development of a real-time aptamer-based sensor that regularly monitors and measures the biomolecules present in the living cells (Ferguson et al. 2013). Moreover, aptamers have also been applied in the diagnosis of pancreatic ductal adenocarcinoma (Wu et al. 2015). Ultrasensitive biosensors have been dispensed for the detection of the virus (Rackus et al. 2015), circulating cancer cells (Green et al. 2016), and nucleic acids (Sekretaryova et al. 2015). Genetically encoded FRET-based nanosensors have been developed for real-time monitoring of flux of different entities like glycine-betaine (Ahmad et al. 2016), methionine (Mohsin and Ahmad 2014), glucose (Chen et al. 2018), and leucine (Mohsin et al. 2016). It utilizes the altered FRET ratio as the measure of the level of specific biomolecules. An improved amyloid detection has also been made possible by the ZnO nanoflower-based biosensor (Akhtar et al. 2017). Recently for glucose diagnosis, a novel frequency-based artificial taste receptor was fabricated with the help of an enzymatic biosensor (Yousefi et al. 2017). Nanobiosensor has also helped in pollution detection in many ways, e.g., cyanide anion detection through the use of Ag@Au core-shell nanoparticle (Dong et al. 2017). Optical biosensing techniques based on the use of surface plasmons are one of the most used forms of biosensing application (Paliwal et al. 2016).

Another biosensor that has gained huge popularity is nanowire biosensor. It is a construction of two extremely active molecules: ssDNA and carbon nanotube serving as detector and transmitter, respectively. Nanowires transmit the chemical event to an output signal in the form of changed conductance. The transmission is reported to be carried out with high sensitivity and in real time and quantitative manner. The properties of nanowires are enhanced by the use of different biological or chemical ligands that help them be analyte independent (Tang et al. 2006). Boron-doped silicon nanowire (SiNWs) is an example of real-time electrically based sensors with high sensitivity for detecting chemical and biological samples. A self-assembling membrane capable of passing electric current has been reported to act as the detecting element for specific molecules and is termed as ion channel switch biosensor. The peculiarity of the sensor lies in the rapid or quick sensing capability that reduces the time detection frame from hours to minutes (Lucas and Harding 2000). Viruses like herpes simplex virus and adenovirus have also been dispensed to use for construction of nanobiosensors with the excellent capability of assembling magnetic nanobeads to be used as sensors (Perez et al. 2003). Probes encapsulated by biologically localized embedding nanobiosensors abbreviated as PEBBLEs have the excellent ability for real-time imaging of inter- and intracellular ions and molecules with null sensitivity to surrounding proteins and bear high stability and reversibility to photobleaching and leaching. PEBBLEs comprise a chemically inert matrix produced by microemulsion polymerization that entraps the sensor molecules used for detection. The sensors are spherical and lie in size range of 20–200 nm (Buck et al. 2004). A powerful oxygen sensing platform provided by PEBBLE has been demonstrated in human plasma, which bears little to no effect by autofluorescence and light scattering (Lee et al. 2009).

The field of application of nanobiosensors encompasses different areas of life including air food and quality (agriculture, climatology, toxins, pollutants, natural disaster), medical sector (diagnosis, pathogen detection, surgical, pharmaceuticals, medical devices), homeland security (air, water, explosives, infectious diseases), health and safety, energy, and quality of life (Sagadevan and Periasamy 2014).

2.2.3.1 Application of FRET-Based Nanobiosensors

Different FRET-based nanobiosensors have been developed in recent years with a wide range of applicability. An overview of some of FRET-based nanobiosensors is given in Table 2.1.

The different areas of applicability of these nanosensors are discussed below:

FRET-Based Nanosensor for Studying Abiotic Stresses

The survival of plants in adverse conditions is mediated through unique growth habits and the different adaptive mechanisms as described by Zhu 2001. One striking feature to overcome high salinity or drought conditions is the accumulation of stress molecules in plants. Many FRET sensors have been developed for understanding the flux of the stress molecules in living cells as described below:

Abscisic Acid

A FRET-based nanosensor named ABACUS was developed to quantify abscisic acid (ABA) which accumulates under adverse conditions. The determination of ABA dynamics was established in prokaryotes, eukaryotes, and Arabidopsis root (Jones et al. 2014). ABA accumulates in the plants to combat adverse conditions and triggers the stomatal closure. The construction of ABACUS included a sensory domain, PAS, which is a member of the ABA co-receptor complexes, and was

Table 2.1 An overview of the developed FRET-based nanosensors

Name of the sensor	Target analyte	LBP employed	FRET pair	Binding affinity (Kd)	Reference
Cameleon YC6.1	Calcium	Calmodulin	CFP-Venus	1130 ± 140 nM	Evanko and Haydon (2005)
FLIPPi	Phosphate	PiBP	ECFP-Venus	840 nM	Gu et al. (2004)
eCALWY-1	Zinc	Atox 1-linker-WD4	Cerulean-Citrine	2 pM	Vinkenborg et al. (2009)
FLIPglu	Glucose	GGBP	CFP-YFP	170 nM	Fehr et al. (2003)
FLIPsuc-4mu	Sucrose	ThuE	ECFP-EYFP	4 µM	Lager et al. (2006)
FLIPmal	Maltose	Male	ECFP-EYFP	2.3 nM	Fehr et al. (2002)
FLIPribose	Ribose	RBP	CFP-YFP	11.7 mM	Lager et al. (2003a, b)
FLIPE	Glutamate	ybeJ	ECFP-Venus	600 nM	Okumoto et al. (2005)
Fluorescent sensor for arginine	Arginine	QBP	ECFP-YFP	2.0 mM	Bogner and Ludewig (2007)
FLIP-Leu	Leucine	LivK	CFP-YFP	192 µM	Mohsin et al. (2013)
FLIPK	Lysine	LAO	CFP-YFP	97 µM	Ameen et al. (2016)

fused to two fluorophores. ABACUS led to the quantitative measurement of ABA with the detection range of ~0.2–800 µM.

Glycine Betaine

The accumulation of glycine betaine is known to occur under conditions of high salinity. It helps in the maintenance of osmolarity of cells without interfering with the physiological processes. A genetically encoded nanosensor based on the principle of FRET named GBOS (Glycine Betaine Optical Sensor) was constructed by sandwiching glycine betaine binding protein (ProX) between cyan fluorescent protein (CFP) and yellow fluorescent protein (YFP). The dynamics of the molecule was studied in both in vitro and in vivo conditions. The expression of the sensor was also carried out in mammalian cells, and the intracellular concentration of the stress molecule was determined (Ahmad et al. 2016).

FRET-Based Nanosensor in Cell Signaing

The complexity and selectivity of signal transduction processes mediated through signaling molecules have been studied through the application of various FRET nanosensors as described below:

Calcium

Calmodulin sandwiched between CFP and Venus was utilized to sense and monitor the cell dynamics of the secondary messenger, Ca^{2+}.The localization of CaM within the ion channel was illustrated through the distance between the donor and acceptor fluorophores. The FRET ratio was noted in both resting cells and Ca^{2+} responsive conditions (Daniel and Philip 2005).

Cyclic Adenosine Monophosphate

The cAMP or cyclic adenosine monophosphate level has been quantified in real time through a FRET nanosensor. The development of the sensor mCerulean-Epac1-mCitrine was achieved through the fusion of Epac1 to mCerulean and mCitrine. The conformational changes in Epac1 reported the concentration change of cAMP in living cells (Salonikidis et al. 2008).

Protein Kinase A Activity

A genetically encoded nanosensor for protein kinase A (PKA) was developed by Zhang et al. (2001). The sensor consisted of a cyan fluorescent protein, a yellow fluorescent protein, and a phosphoamino acid binding domain (14–3-3τ), a consensus substrate for PKA. Phosphorylation led to structural changes which altered the yellow/cyan emission intensities by 25–50% in living cells. The tethering of the sensor protein to PKA holoenzyme and the localization of the sensor to the nucleus led to accelerated and decelerated FRET response, respectively. The study concluded that colocalization of a substrate and PKA or redistribution of a substrate can affect or control the susceptibility of the substrate to kinase phosphorylation. The study laid a base for studying the compartmentation and targeting of PKA and different kinases and phosphatases.

In another study, visualization of protein phosphorylation in living cells through a fluorescent nanosensor was reported (carried out by Nagai et al. 2000). The nanosensor consisted of two GFP variants linked to the kinase-inducible domain (KID) of CREB (transcription factor of cyclic adenosine monophosphate (cAMP)-responsive element binding protein). The transfection of COS-7 cells with the sensor protein led to the visualization of PKA activation dynamics. The phosphorylation of KID mediated by the cAMP-dependent PKA decreased the FRET intensity between the GFPs.

Tyrosine Phosphorylation

A nanosensor based on FRET for monitoring the level of CrkII tyrosine phosphorylation, an adaptor protein involved in the signaling events such as apoptosis, migration, and growth in living cells, was developed (Kurokawa et al. 2001). The sensor involved the fusion of CrkII to CFP and YFP on either end. The transient and rapid CrkII phosphorylation triggered through epidermal growth factor was monitored in single cells. The site of initiation of phosphorylation activity was also revealed by the fusion of CAA*X* box of Ki-Ras to the C-terminus of the nanosensor.

FRET-Based Nanosensor in Therapeutics

The flux study of the compounds or molecules involved in different diseases like diabetes, cancer, and neurodegeneration has been elucidated through the development of FRET-based nanosensors as described below.

Glucose

The quantification of glucose in living cells was determined by a nanosensor comprising of two GFP variants: CFP and YFP ligated at the N- and C-terminus of the glucose-galactose binding protein (GGBP). The binding affinity constant of the protein was calculated as 170 nM and the sensor was named as FLIPglu. A decrease in FRET ratio was recorded upon the binding of glucose to GGBP. The maximum FRET ratio change was recorded as 0.23. Different affinity mutants of the wild-type sensor were also developed, and the dynamics of glucose in COS-7 cells was monitored (Fehr et al. 2003).

Vitamin B12

The essentially important vitamin, vitamin B12, has an important role in the metabolism of human cell. The flux dynamics of vitamin B12 or cobalamin was studied through a FRET sensor specific for vitamin B12 (Ahmad et al. 2018). A construct for the nanosensor was designed by using BtuF (vitamin B12 binding protein) as a recognition element and CFP and YFP as reporter proteins. The sensor was shuttled in different expression vectors, and the purified protein was named SenVitAL (Sensor for Vitamin Anemia Linked). The sensor was found specific for vitamin B12 and pH stable. The K_d of the protein was calculated as ~157 μM. It allowed the measurement of vitamin B12 from micro-molar to milli-molar levels. An increased FRET ratio increase was observed upon addition of vitamin B12 during in vitro and in vivo analysis. The dynamics of vitamin B12 was also measured in yeast and mammalian cells.

BCR-ABL Kinase

For the assessment of drug efficacy utilized for chronic myeloid leukemia (CML) patients, evaluation of BCR-ABL kinase activity in living cells was carried out using a FRET-based biosensor. A characteristic substrate of BCR-ABL, CrkL, was used as the sensory domain, sandwiched between Venus and ECFP. An increased FRET efficiency was recorded due to the binding of the SH2 domain of CrkL to phosphorylated tyrosine (Y207). The comparison of the biosensor properties with established methods like flow cytometry and western blotting and the activity of BCR-ABL in response to drugs were examined in the cells of CML patients (Mizutani et al. 2010).

Lactate

FRET nanosensor developed by San Martín et al. (2013) enabled the monitoring of lactate flux between and within the living cells. The lactate production is related to disrupted metabolism, and its involvement is also known in hypoxia/ischemia, cancer, inflammation, and neurodegeneration. mTFP and Venus were attached to the N- and C-termini of lactate binding bacterial transcriptional protein (LldR) and the sensor protein was named as Laconic. The quantification of lactate levels was enabled by Laconic in the range of 1–10 mM.

FRET-Based Nanosensor in Industries

The study of live cell dynamics of industrially important biomolecules like amino acids has also been carried out through FRET sensors.

Methionine

Mohsin and Ahmad (2014) developed a nanosensor based on FRET principle for studying the in vivo dynamics of methionine in living cells. The methionine binding protein (MetN) was amplified from *E. coli* K12 and used as the reporter element. The sensor was constructed by the fusion of CFP and YFP at the N- and C- termini of MetN, respectively. The sensor was expressed, and the FRET changes were recorded in both bacterial and yeast cells. The sensor developed was named FLIPM, and the K_d calculated was 203 mM. The sensor was found pH stable in the physiological range and highly specific for methionine.

Leucine

In 2013, Mohsin and coworkers studied the dynamics of leucine in living cells. The study was undertaken due to the vital importance of leucine in livestock industry. Leucine binding protein (LivK) amplified from *E. coli* was used as the recognition element in the construction of the sensor. The fluorophores used to report the changes in the level of leucine were CFP and YFP. The developed sensor was named FLIP-Leu, and the Kd calculated for the sensor was 192 μM. FLIP-Leu helped in the measurement of lysine in both *in vitro* and in vivo conditions in real time.

Glutamate

The glutamate binding protein, ybeJ, was used as the sensing element in the construction of glutamate sensor, FLIPE. ybeJ was fused with two fluorescent proteins, Venus and ECFP, on its either end. FRET ratio change was recorded with the change in the glutamate concentration. In addition, different affinity mutants were also developed through point mutation in the ligand binding domain of ybeJ. The Kd of FLIPE was calculated as 600 nM (Okumoto et al. 2005).

Lysine

The in vivo monitoring of the industrially important amino acid, lysine, was carried out by Ameen et al. (2016). A periplasmic binding protein specific for lysine (LAO) was isolated from *Salmonella typhimurium* LT2 strain and used as the reporter element. The N- and C-termini of LAO protein were fused with CFP and YFP to produce the sensor protein (FLIPK). In vitro characterization of the protein was carried out wherein the pH stability, specificity, affinity, and metal ion effects were analyzed. Affinity of FLIPK was calculated as 97 μM, and the pH stability of the protein was found in the physiological range. Moreover, affinity mutants of the sensor were also developed and investigated. The study helped in the real-time monitoring of lysine dynamics in prokaryotic as well as eukaryotic cells.

2.2.4 Mechanical Sensors

Mechanical sensors structure a class of sensors that are sensitive to changes in mechanical properties. In the previous decade, different mechanical sensors with various components have been illustrated. The earliest mechanical nanosensor was projected by Binh et al. 1994 for estimating the undulation and flexible attributes of a nanosphere attached to a tapered cantilever. This work is imperative for application in nano-devices segments and nanoscale subassemblies in microelectronic devices. Rather than estimating the undulation attributes and elastic properties of the subassemblies joined to a surface, Binh et al. 1994 gave the idea of manufacturing replicas of these articles from the heating of fine wires terminated with sharp tips.

Cantilevers and acoustic sensors are two well-known and popular mechanical sensors that assume an essential job in the molecular study (Zhang and Hoshino 2018). Cantilevers are usually rectangular-formed bars of Si under 1 μm thick. Adsorption/acknowledgment of particles on the surface of such micromechanical cantilevers functionalized with receptor atoms brings about bowing of the cantilever as a result of the surface pressure (Lang et al. 2005). It has been incontestable that cantilever biosensors offer fascinating conceivable outcomes for label-free detection of biomolecules. Here, the nanomechanical bowing of the cantilever is prompted by a differential surface stress due to the molecular binding to its functionalized surface. (Zhang and Hoshino 2018). These strategies have been utilized as general techniques for identifying different sorts of biomolecules (Mukhopadhyay et al. 2005).

Acoustic wave sensor works by checking the adjustment in the physical properties of an acoustic wave. The acoustic sensor utilizes piezoelectric materials to create acoustic waves in any solid materials using properly customized electric fields, and further acoustic waves are detected due to charge produced by the incited mechanical distortion. Quartz is regularly utilized as it is copious, is amiable to minimal effort volume fabrication, and is related with unusual mechanical properties and great concoction dependability. Contingent upon the cut of the precious quartz stone, diverse features can be figured out. There are two customary "cuts" in acoustic sensors: the AT cut and the ST cut. These distinctions cause diverse piezoelectric disfigurements, which enable acoustic waves to spread in various ways productively, along the outside of the sensor or far from the sensor surface toward the mass substrate. This qualification is very critical and is the premise of the order of acoustic sensors into either surface acoustic wave (SAW) or mass acoustic wave (BAW) sensors (Fogel et al. 2016). These sensors are inherently sensitive and are fit for analyzing different physical parameters.

These gadgets can likewise be scaled down with focal points regarding cost, size and versatility, scaled sensitivities related to smaller measurements, and higher operational frequencies. The capacity to multiplex recognition crosswise over varieties of many gadgets installed in a solitary chip expanded throughput and the capacity to investigate a more extensive scope of modes including inside a similar gadget. Furthermore, device fabrication is regularly perfect with semiconductor volume batch manufacturing empowering cost versatility and a high level of preciseness and reproducibility in the assembling procedure. Microfluidic integration can also lead to enhance selectivity and overall signal to voice ratio.

In the last couple of years, miniaturization has received a lot of attention as it had led to the radical enhancement of the sensor performance. Parallel to this advancement, microelectromechanical system (MEMS) has supplanted mechanical sensors in many applications. In the simplest way, MEMS can be explained as nothing but tiny devices consisting of both electrical and mechanical components. Due to the possibility of fabrication of MEMS using semiconductor device fabrication technologies, it became easy to employ MEMS as a sensor practically. MEMS innovation has been exceptionally fruitful in the physical detecting setting and has yielded a scope of small, rough, and cheap gadgets. One of the most energizing and dynamic aspects of MEMS is its application in biomedical and health science, which involves the integration of biological detecting capacities and materials into a MEMS device. Such MEMS are referred to as BioMEMS – biological microelectromechanical systems. A standout among the greatest BioMEMS applications is the miniaturized PCR (polymerase chain response) framework, and the model is the in-check item created by electronics giant ST Microelectronics. This incorporates the majority of the capacities expected to distinguish a given oligonucleotide arrangement and incorporates the microfluidic dealing with the framework, i.e., a miniaturized PCR reactor and a custom microarray.

2.2.5 Optical Sensors

Because of the potential for giving an enhanced understanding of cell responses to various stimuli, a lot of interest has been devoted to the subject of optical nanosensors. An optical nanosensor is a measuring gadget which comprises two parts: a bioreceptor and a transducer. The bioreceptor is a biomolecule that perceives the target analyte, and this perception event changes into quantifiable optical signals by the transducer. The uniqueness of optical nanosensors is that the two segments are incorporated into one single sensor.

Figure 2.1 demonstrates the essential guideline of optical nanosensors. This mix empowers one to gauge the objective analyte without utilizing reagents. For instance, the benzo(a)Pyrene (BaP) can be explicitly estimated inside a solitary living cell by an optical nanosensor which is made explicitly for BaP estimation by embedding the nanosensor in the cell. This is as opposed to the common measure in which various means are utilized, and each progression may require a reagent to treat the cells. The effortlessness and the speed of estimation are the principles of favorable circumstances of optical nanosensors. One noteworthy prerequisite for optical nanosensors is that the bioreceptor particle must be immobilized in direct spatial coupling to the transducer. The immobilization is done either by physical entanglement or by chemical attachment. The first optical nanosensor was reported

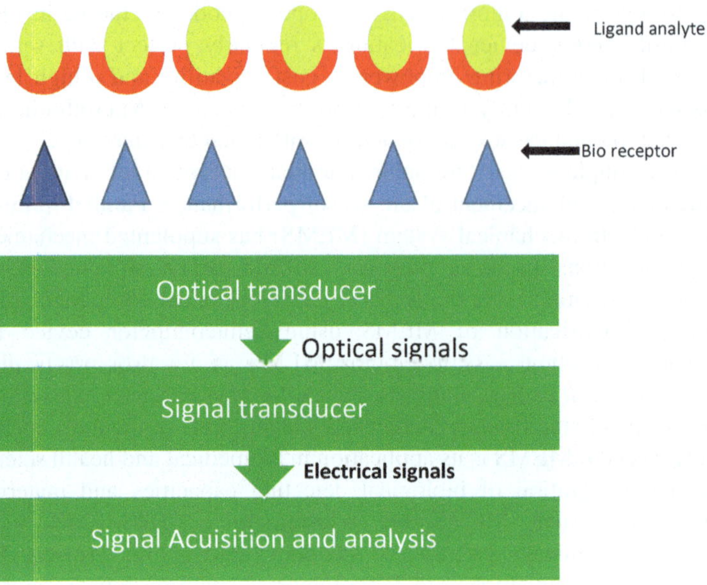

Fig. 2.1 The essential guidelines of optical nanosensors

for pH measurement and was based on fluorescein (De Silva et al. 1997; Czarnik 1993). Fluorescein is a fluorophore commonly used in microscopy. Fluorescent chemosensors exploit luminescence phenomenon which involves absorption of light of certain wavelength by fluorophore followed by excitation and emission of a quantum of light with an energy corresponding to the energy difference between the ground and excited states (Kulmala and Suomi 2003). The sensing concept of this sensor relies on the changes in photo-vibrational properties. The most basic type of optical nanosensor is that of the molecular dye probe (Haugland 1996) within a cell that involves immediate cell loading of fluorescent dyes. A plus point of this approach is that it reduces the physical disturbances of the cell, which is in contrast to the optical fiber probe. However, an obstacle of the free dye method is the inherent dye-cell chemical interference (Lim and Ramakrishna 2006). An extraordinary marginal deviation from the free dye method is the tagged nanoparticle that consists of a reporter molecule attached to the surface of the nanoparticle (Ji et al. 2000; Ji et al. 2001).

2.2.5.1 Fiber-Optic Nanosensor

Conventional strategies for intracellular examination need fixing of the cell before carrying out the investigation which decimates cell suitability and change the intracellular structure to a remarkable degree. The most broadly announced class of optical nanosensors is fiber-optic nanosensors, which can carry out the analysis in vivo (Lim and Ramakrishna 2006). Tan and coworkers firstly contributed toward the development of fiber-optic nanosensor [Tan et al. 1992 (a); Tan et al. 1992 (b)].

Fundamental processes at the cell level are imperative to improve further understanding of dynamic cell capacities which can be easily achieved by fiber-optic nanosensor as they have the potential to scrutinize necessary cellular processes in vivo. When target molecule reacts with the receptor or experiences a physiochemical perturbation, the change can be converted to electrical signal or any other quantifiable signals [Cullum et al. 2000; Kasili et al. 2002; Vo-Dinh et al. 2000(a); Vo-Dinh et al. 2000(b); Vo-Dinh et al. 2000(c)].

R + A → RA + measurable signal (Binh et al. 1994).

These quantifiable signals are then sensed by the optical probe and transmitted into the database. The limitations of the dye-cell compound impedance are overcome by the free dye technique due to the utilization of the optical fiber probe which minimizes the effect of the environmental perturbations on the detecting tip. Another favorable position of the optical nanosensor is the insignificant obtrusiveness of this method as compared with customary wire-probe devices. Fiber-optic nanosensors have so far been effective with their capacity in the applications like measurement of benzopyrene tetrol (BPT) and benzo[*a*]pyrene (BaP) inside single cells (Cullum et al. 2000; Kasili et al. 2002; Vo-Dinh et al. 2000(a); Vo-Dinh et al. 2000(b); Vo-Dinh et al. 2000(c)), monitoring apoptosis, measuring caspase-9, measuring cytochrome c (Song et al. 2004), and so forth. However, because of the little inspecting volume examined by the optical fiber nanosensor, the measure of a target analyte in the excitation volume is small, consequently making it necessary to receive a delicate optical spectroscopic strategy (e.g., fluorescence) for investigation (revue naZ).

2.2.5.2 PEBBLEs

The photonic explorers for bio-analysis with biologically localized embedding (PEBBLE) was presented to defeat the limitations related to both the free fluorescent dye and the optical fiber technique. These are nanoscale detecting gadgets which epitomize an analyte-explicit dye and a reference dye within a biologically inert matrix (Clark et al. 1999). PEBBLEs are less physically troublesome to the cellular condition due to the absence of an extended probe associating the sensor in the cell with the outside of the cell. Besides, the embodiment of the dye within the inert matrix guarantees that the sensing phase is isolated from the cell setting and, hence, forestalling chemical impedance. PEBBLEs can be sorted into four kinds as per their particular framework, which are two classifications based on their working standards, and four techniques for PEBBLE conveyance into the cell. The four kinds of PEBBLE matrices are:

1. Polyacrylamide PEBBLEs are made by polymerizing a solution of monomer, sensing dye, reference dye, and a surfactant to control the size.
2. Polymethylmethacrylate (PDMA) PEBBLEs are polymerized inside a hydrophobic condition without any dye molecule or any other sensing element (Brasuel et al. 2001; Brasuel et al. 2002a, b; Brasuel et al. 2003).
3. Sol-gel PEBBLEs are integrated utilizing "delicate" systems that permit the consideration of sensitive biological molecules. The sol-gel nanoparticle planning

is completed within sight of the detecting segments to improve biocompatibility as these PEBBLEs are covered with poly(ethylene glycol) (Xu et al. 2001).

4. Organically modified silicates (Ormosils) PEBBLEs are set up in two stages. In the main stage, the center development starts by hydrolyzing phenyltrimethoxysilane inside the acidic condition, pursued by silane condensation within the alkalescent state. Later the nanoparticle canters are coated with the ormosil layer. And then lastly, the detecting components are consolidated into the ormosil PEBBLEs just before the second layer shapes (Koo et al. 2004).

Further based on the working standard of PEBBLEs, two types have been identified: (a) direct measurement PEBBLEs and (b) ion correlation.

The direct measurement PEBBLEs are used for detecting both ions and small molecule. These permit the analyte to saturate the matrix and interface with the indicator dye directly and specifically causing fluorescence incitement or fluorescence quenching. However, some analytes need exceedingly specific fluorescent markers. Ion correlation or ion exchange PEBBLEs have been created to address this inadequacy (Lim and Ramakrishna 2006). Gene gun, pico injection, liposomal, and cell-directed are the four methods of PEBBLEs delivery (Brasuel et al. 2002a, b).

2.2.5.3 Quantum Dots

The distinctive optical properties of quantum dots (QDs) make them desirable as in vivo and in vitro fluorophores in an assortment of biological investigation where fluorescent label based on organic molecule misses the mark regarding giving stability and simultaneous signal detection. The capacity to make QDs water solvent and target them to specific biomolecules has prompted promising applications in cell labeling, tissue imaging, and assay labeling and as productive fluorescence energy transfer donors (Medintz et al. 2005). QDs exhibit an important feature of having a broad absorption band, which further allows for the selection of diverse possible excitation wavelength from visible to near-infrared regions. They are also assumed to be superior in terms of luminescence lifetimes, resistance against photobleaching, and narrow emission band because of which they are considered as an ideal choice for versatile designing of the sensor (Cui et al. 2015). A large number of quantum chemosensors have been developed for transition metal ions and small molecules by utilizing their properties. Quantum dots have opened up a new research area where it is employed for the detection of biomolecules like protein, DNA, and saccharides. The QDs associated with a ligand–quencher conjugate demonstrate a turn-on PL upon the association of the conjugate with the test biomolecule. Advanced QD sensors with a surface-immobilized redox focus have to been created to tune the PL of QDs by electron transfer enabling the photoluminogenic testing of biomolecules on both the atomic and cell levels. On account of their high restricting proclivity and specificity for DNA, organometallic mixes were utilized for conjugation. By exploiting their predominant optical properties counting high signal brilliance and photostability, biofunctionalized QDs have been utilized for cell imaging. QDs brightened with a focusing on the operator that can encourage the take-up of the QD by specific cells or potentially with a responsive site with the capacity to connect with a defined intracellular species have been extensively prepared for the "target-specific" imaging of live cells (Cui et al. 2015).

2.2.6 Thermal Sensors

Temperature is the proportion of the typical active vitality of the atoms of a gas, fluid, or solid. It is a gadget that is explicitly used to measure temperature. Along these lines, warm sensors can give us a quantifiable method to depict the substance, regardless of whether it is an article nature in which an item is set or the earth in which an item is conveyed. Thermal sensors also called as temperature sensors are constructed for measuring temperature changes. The very first and popular thermal sensor still in use is a thermometer. It consists of either mercury or alcohol whose level rises on increasing the temperature, hence indicating the temperature fluctuation. However, such type of sensor cannot be suited on a wider scale. Thus, for the field of electronics and microprocessor, there are other thermal sensors that are being employed.

1. *Thermocouple*: It involves the coupling of two different metals at their detecting end. Once there is a gradient between the recent device element and cold reference junction, a voltage is created, and the amendment or the change in this voltage can be accounted for as temperature (Love 2007). Figure 2.2.gives a rough idea about thermocouple construction.
2. *Resistance thermometer*: Resistance thermometer or resistance temperature detectors or RTDs are usually made up of single pure metal having the property of electrical obstruction or resistance, which is nothing but the function of temperature. Platinum is the most precise metal being used for resistance thermometer as it has an exceptionally linear relationship with the temperature.
3. *Thermistors*: These are a type of resistance thermometers that are created by connecting metal wires to a ceramic base made of several sintered, oxide semiconductors (Janata 2010). These are similar to resistance thermometers as the change in resistance corresponds to change in temperature. However, the relationship it shares with temperature is not very linear because of which it covers a narrow range of temperature. Due to its small size, inexpensiveness, and sensitiveness to temperature, they are ideally suited for electronic applications.

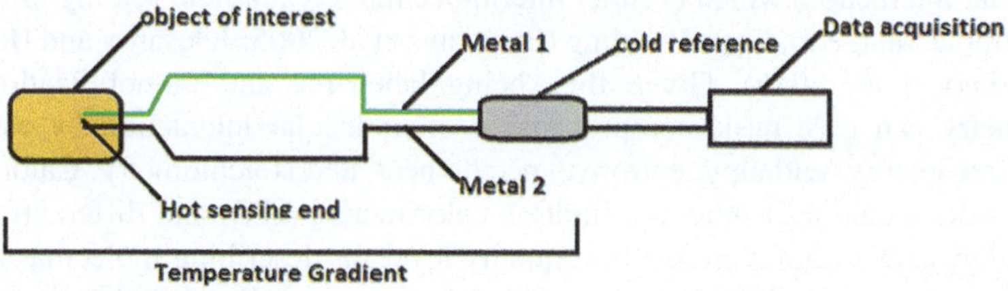

Fig. 2.2 Thermocouple construction

4. *Silicon sensors*: It is made up of silicon which is a base material for most of the electronic microchips. The assemblage of these electronic gadgets is a cautiously controlled, high-volume fabricating process that incorporates doping, and cautious layering of metals, oxides, and encasings (Peterson 1983).
5. *Radiation thermometers*: All substances and articles emanate thermal radiation when it is at a temperature higher than total zero (0 K or − 273.15 °C). There is a connection between temperature and radiation vitality discharged that can be utilized to figure the temperature of the item surface. In contrast to all abovementioned sensors, radiation thermometers are principally utilized at a separation from the object of intrigue and can be utilized for difficult-to-achieve objects.

2.2.6.1 Thermal Sensor for a Biological System

Thermogenesis is an important aspect of living organisms which can be defined as the production of heat in a human or animal body. In order to study thermogenesis and temperature dissemination at the single-cell level, there is a need of sensors that are insulated and protected and are equipped for examining temperature conveyances in sub-micrometer zones without meddling with any ecological parameter in the surrounding medium (ElShimy et al. 2007).Ultra-small thermal sensors are developed by applying distinctive methodologies which incorporate temperature sensors dependent on carbon nanotubes (CNT), nuclear power magnifying lens, cantilever-based temperature sensor, thermosensitive fluorescent colors, and indium tin oxide (ITO) microchips. Novel nanotools like electrostatic controller (Gu 2008; Kometani et al. 2004), the bio-nano-injector (Kometani et al. 2003), nano-nets (which are utilized in manipulating subcell organelles) are fabricated for high exactitude with FIB-CVD in recent years. Thermogenesis is not only limited in living organism but can also be found effective in physical or biological reaction or chemical reaction as many chemical or physical reactions catalyzed by enzyme are exothermic, creating heat which might be measured by calorimeter and further can be utilized as a reason for estimating the rate of response and, subsequently, the analyte fixation. Calorimeter employs the usage of thermistors, and hence, it is an important aspect of thermal nanosensor. The calorimeter is not only used for heat measurement of exothermic reaction but can also be used to study bimolecular interactions. All organic processes depend somehow on atomic collaboration and molecular interaction, which is either intermolecular, e.g., protein folding/unfolding, or intermolecular, e.g., ligand binding (Bruylants et al. 2005; Jelesarov and Bosshard 1999; Zuo et al. 2016). Given this, being label-free and immobilization free, calorimetry can give basic comprehension of molecular interaction by assessing Gibbs free energy, enthalpy, entropy, specific heat, and stoichiometry. Calorimetric frameworks including isothermal titration calorimeters (ITC) and differential scanning calorimeters (DSC) are the best quality level for describing molecular interaction (Bruylants et al. 2005; Chaires 2008; Holdgate and Ward 2005; Wang et al. 2017; Zuo et al. 2016). The noteworthiness of the calorimeters in the organic investigation has persuaded scientists to limit them in the previous decade, including scaling down gadgets, i.e., miniaturizing and lessening test consumption. Nowadays, studies are being done on small-scale creation of a power remuneration differential scanning nano-calorimeter with the point of diminishing the sample volume to 1–5 μl and enhancing the temperature affectability to 10 μK, which compare to the prerequisites of typical pharmaceutical medication screening or a weak binder study. The nano-calorimeter comprises four thermistors. Indistinct SiC is picked, and the creative state of DC magnetron sputtering is being optimized to get susceptible temperature detecting material (Zuo et al. 2016).

2.3 Conclusion

As discussed in the chapter, the extreme usefulness of nanosensors deployed in the different field of research can be gathered to a great deal of understanding. The application of the developed nanosensors has reached a milestone in environmental, pharmaceutical, and medical fields and have the paved way for the fabrications of even more useful sensing devices. The nanotechnological advances in the coming years will help in unraveling the present enigma in the field of biology. The present challenge faced by the researchers is the reliable use of the techniques in the manufacturing of working nanoscale devices. The shrinkage of devices is related to hindrances in the form of its melting point, chemical reactivity, or electrical conductivity, which can alter its performance. Therefore, the need to master the technology should be the concern of the scientists across the nations, which will generate opportunities to improve the areas of biological life.

Acknowledgments The first author (GA) is thankful to the Council of Scientific and Industrial Research, Govt. of India, for providing Senior Research Fellowship.

References

Ahmad M, Ameen S, Siddiqi TO, Khan P, Ahmad A (2016) Live cell monitoring of glycine betaine by FRET-based genetically encoded nanosensor. Biosens Bioelectron 86:169–175

Ahmad M, Mohsin M, Iqrar S, Manzoor O, Siddiqi TO, Ahmad A (2018) Live cell imaging of vitamin B12 dynamics by genetically encoded fluorescent nanosensor. Sensor Actuat B Chem 257:866–874

Akhtar N, Metkar SK, Girigoswami A, Girigoswami K (2017) ZnO nanoflower based sensitive nano-biosensor for amyloid detection. Mater Sci Eng C 78:960–968

Ameen S, Ahmad M, Mohsin M, Qureshi MI, Ibrahim MM, Abdin MZ, Ahmad A (2016) Designing, construction and characterization of genetically encoded FRET-based nanosensor for real time monitoring of lysine flux in living cells. J Nanobiotechnol 14:49

Bellan LM, Wu D, Langer RS (2011) Current trends in nanobiosensor technology. Wiley Interdisciplinary Reviews: Nanomed Nanotechnol 3:229–246

Binh VT, Garcia N, Levanuyk AL (1994) A mechanical nanosensor in the gigahertz range: where mechanics meets electronics. Surf Sci 301:L224–L228

Blind M, Blank M (2015) Aptamer selection technology and recent advances. Mol Ther Nucleic Acids 4:e223

Bogner M, Ludewig U (2007) Visualization of arginine influx into plant cells using a specific FRET-sensor. J Fluoresc 17:350–360

Brasuel M, Kopelman R, Aylott JW, Clark H, Xu H, Hoyer M, Miller TJ, Tjalkens R, Philbert MA (2002a) Production, characteristics and applications of fluorescent PEBBLE nanosensors: potassium, oxygen, calcium and pH imaging inside live cells. Sensors Mater 14:309–338

Brasuel M, Kopelman R, Kasman I, Miller TJ, Philbert MA (2002b) Ion concentrations in live cells from highly selective ion correlation fluorescent nano-sensors for sodium. In: Sensors, vol 1. IEEE, Piscataway, pp 288–292

Brasuel M, Kopelman R, Miller TJ, Tjalkens R, Philbert MA (2001) Fluorescent nanosensors for intracellular chemical analysis: decyl methacrylate liquid polymer matrix and ion-exchange-based potassium PEBBLE sensors with real-time application to viable rat C6 glioma cells. Anal Chem 73:2221–2228

Brasuel MG, Miller TJ, Kopelman R, Philbert MA (2003) Liquid polymer nano-PEBBLEs for cl– analysis and biological applications. Analyst 128:1262–1267

Brown SD, Bear RS, Blank TB (1992) Chemometrics. Anal Chem 64:PR22–PR49

Bruylants G, Wouters J, Michaux C (2005) Differential scanning calorimetry in life science: thermodynamics, stability, molecular recognition and application in drug design. Curr Med Chem 12:2011–2020

Buck SM, Xu H, Brasuel M, Philbert MA, Kopelman R (2004) Nanoscale probes encapsulated by biologically localized embedding (PEBBLEs) for ion sensing and imaging in live cells. Talanta 63:41–59

Chaires JB (2008) Calorimetry and thermodynamics in drug design. Annu Rev Biophys 37:135–151

Chandra P, Noh HB, Pallela R, Shim YB (2015) Ultrasensitive detection of drug resistant cancer cells in biological matrixes using an amperometric nanobiosensor. Biosens Bioelectron 70:418–425

Chen L, Hwang E, Zhang J (2018) Fluorescent Nanobiosensors for sensing glucose. Sensors 18:1440

Clark HA, Hoyer M, Philbert MA, Kopelman R (1999) Optical nanosensors for chemical analysis inside single living cells. 1. Fabrication, characterization, and methods for intracellular delivery of PEBBLE sensors. Anal Chem 71:4831–4836

Cooper MA (ed) (2009) Label-free biosensors: techniques and applications. Cambridge University Press, Cambridge

Cui L, He XP, Chen GR (2015) Recent progress in quantum dot based sensors. RSC Adv 5:26644–26653

Cullum BM, Griffin GD, Miller GH, Vo-Dinh T (2000) Intracellular measurements in mammary carcinoma cells using fiber-optic nanosensors. Anal Biochem 277:25–32

Cusano A, Giordano M, Cutolo A, Pisco M, Consales M (2008) Integrated development of chemoptical fiber nanosensors. Curr Anal Chem 4:296–315

Czarnik AW (ed) (1993) Fluorescent chemosensors for ion and molecule recognition. American Chemical Society, Washington

da Silva ETSG, Miserere S, Kubota LT, Merkoçi A (2014) Simple on-plastic/paper inkjet-printed solid-state Ag/AgCl pseudoreference electrode. Anal Chem 86:10531–10534

Daniel SE, Philip GH (2005) Elimination of environmental sensitivity in a cameleon FRET-based calcium sensor via replacement of the acceptor with Venus. Cell Calcium 37:341–348

de la Escosura-Muñiz A, Merkoçi A (2016) Nanochannels for electrical biosensing. Trac Trend Anal Chem 79:134–150

De Palma R, Reekmans G, Liu C, Wirix-Speetjens R, Laureyn W, Nilsson O, Lagae L (2007a) Magnetic bead sensing platform for the detection of proteins. Anal Chem 79:8669–8677

De Palma R, Reekmans G, Liu C, Wirix-Speetjens R, Laureyn W, Nilsson O, Lagae L (2007b) Magnetic bead sensing platform for the detection of proteins. Anal Chem 79:8669–8677

De Silva AP, Gunaratne HN, Gunnlaugsson T, Huxley AJ, McCoy CP, Rademacher JT, Rice TE (1997) Signaling recognition events with fluorescent sensors and switches. Chem Rev 97:1515–1566

Dong ZZ, Yang C, Vellaisamy K, Li G, Leung CHand Ma DL (2017) Construction of a nano biosensor for cyanide anion detection and its application in environmental and biological systems. ACS Sens 2:1517–1522

ElShimy HM, Nakajima M, Arai F, Fukuda T (2007) Nano thermal sensors for sensing temperature in water environment. In: 2007 7th IEEE conference on nanotechnology (IEEE NANO). IEEE, Piscataway, pp 1045–1049

Evanko DS, Haydon PG (2005) Elimination of environmental sensitivity in a cameleon FRET-based calcium sensor via replacement of the acceptor with Venus. Cell Calcium 37:341–348

Fehr M, Frommer WB, Lalonde S (2002) Visualization of maltose uptake in living yeast cells by fluorescent nanosensors. Proc Natl Acad Sci U S A 99:9846–9851

Fehr M, Lalonde S, Lager I, Wolff MW, Frommer WB (2003) In vivo imaging of the dynamics of glucose uptake in the cytosol of COS-7 cells by fluorescent nanosensors. J Biol Chem 278:19127–19133

Ferguson BS, Hoggarth DA, Maliniak D, Ploense K, White RJ, Woodward N, Hsieh K, Bonham AJ, Eisenstein M, Kippin TE, Plaxco KW (2013) Real-time, aptamer-based tracking of circulating therapeutic agents in living animals. Sci Transl 5:213ra165

Fogel R, Limson J, Seshia AA (2016) Acoustic biosensors. Essays Biochem 60:101–110

Grate JW, Rose-Pehrsson SL, Venezky DL, Klusty M, Wohltjen H (1993) Smart sensor system for trace organophosphorus and organosulfur vapor detection employing a temperature-controlled array of surface acoustic wave sensors, automated sample preconcentration, and pattern recognition. Anal Chem 65:1868–1881

Green BJ, Safaei TS, Mepham A, Labib M, Mohamadi RM, Kelley SO (2016) Beyond the capture of circulating tumor cells: next-generation devices and materials. Angew Chem Int Ed 55:1252–1265

Grossman HL, Myers WR, Vreeland VJ, Bruehl R, Alper MD, Bertozzi CR, Clarke J (2004) Detection of bacteria in suspension by using a superconducting quantum interference device. Proc Natl Acad Sci U S A 101:129–134

Gu W (2008) Modeling and application of flow-limited field-injection electrostatic spraying (FFESS). University of Illinois at Urbana-Champaign, Urbana

Gu Y, Di WL, Kelsell DP, Zicha D (2004) Quantitative fluorescence resonance energy transfer (FRET) measurement with acceptor photobleaching and spectral unmixing. J Microsc 215:162–173

Haswell SJ (ed) (1992) Practical guide to chemetrics. M. Dekker, New York

Haugland RP (ed) (1996) Handbook of fluorescent probes and research chemicals. Molecular Probes, Eugene

Hirsch A, Vostrowsky O (2005) Functionalization of carbon nanotubes. In: Schluter AD (ed) Functional molecular nanostructures. Heidelberg, Springer, pp 193–237

Holdgate GA, Ward WH (2005) Measurements of binding thermodynamics in drug discovery. Drug Discov Today 10:1543–1550

Hong CY, Chen WS, Jian ZF, Yang SY, Horng HE, Yang LC, Yang HC (2007) Wash-free immunomagnetic detection for serum through magnetic susceptibility reduction. App Phys Lett 90:074105

Hong CY, WuCC CYC, Yang SY, Horng HE, Yang HC (2006) Magnetic susceptibility reduction method for magnetically labeled immunoassay. App Phys Lett 88:212512

Howarter JA, Youngblood JP (2006) Optimization of silica silanization by 3-aminopropyltriethoxysilane. Langmuir 22:11142–11147

Ikebukuro K, Kiyohara C, Sode K (2005) Novel electrochemical sensor system for protein using the aptamers in sandwich manner. Biosens Bioelectron 20:2168–2172

Janata J (2010) Principles of chemical sensors. Springer Science & Business Media, London

Jelesarov I, Bosshard HR (1999) Isothermal titration calorimetry and differential scanning calorimetry as complementary tools to investigate the energetics of biomolecular recognition. J Mol Recognit 12:3–18
Ji J, Rosenzweig N, Jones I, Rosenzweig Z (2001) Molecular oxygen-sensitive fluorescent lipobeads for intracellular oxygen measurements in murine macrophages. Anal Chem 73:3521–3527
Ji J, Rosenzweig N, Griffin C, Rosenzweig Z (2000) Synthesis and application of submicrometer fluorescence sensing particles for lysosomal pH measurements in murine macrophages. Anal Chem 72:3497–3503
Jianrong C, Yuqing M, Nongyue H, Xiaohua W, Sijiao L (2004) Nanotechnology and biosensors. Biotechnol Adv 22:505–518
Jones AM, Danielson JÅ, Manoj Kumar SN, Lanquar V, Grossmann G, Frommer WB (2014) Abscisic acid dynamics in roots detected with genetically encoded FRET sensors. elife 3:e01741
Josephson L, Perez JM, Weissleder R (2001) Magnetic nanosensors for the detection of oligonucleotide sequences. Angew Chemie (International ed In English) 40:3204–3206
Kashefi-Kheyrabadi L, Mehrgardi MA, Wiechec E, Turner AP, Tiwari A (2014) Ultrasensitive detection of human liver hepatocellular carcinoma cells using a label-free aptasensor. Anal Chem 86:4956–4960
Kasili PM, Cullum BM, Griffin GD, Vo-Dinh T (2002) Nanosensor for in vivo measurement of the carcinogen benzo [a] pyrene in a single cell. J Nanosci Nanotechnol 2:653–658
Kissinger P (2005) Biosensors – a perspective. Biosens Bioelectron 20:2512–2516
Koh I, Josephson L (2009) Magnetic Nanoparticle Sensors. Sensors 9:8130–8145
Koh I, Hong R, Weissleder R, Josephson L (2009) Nanoparticle-target interactions parallel antibody-protein interactions. Anal Chem 81:3618–3622
Kometani R, Hoshino T, Kondo K, Kanda K, Haruyama Y, Kaito T, Fujita JI, Ishida M, Ochiai Y, Matsui S (2004) Characteristics of Nano-electrostatic actuator fabricated by focused ion beam chemical vapour deposition. Jpn J Appl Phys 43:7187
Kometani R, Morita T, Watanabe K, Kanda K, Haruyama Y, Kaito T, Fujita J, Ishida M, Ochiai Y, Matsui S (2003) Nozzle-nanostructure fabrication on glass capillary by focused-ion-beam chemical vapor deposition and etching. Jpn J App Phys 42:4107
Koo YEL, Cao Y, Kopelman R, Koo SM, Brasuel M, Philbert MA (2004) Real-time measurements of dissolved oxygen inside live cells by organically modified silicate fluorescent nanosensors. Anal Chem 76:2498–2505
Kulmala S, Suomi J (2003) Current status of modern analytical luminescence methods. Anal Chim Acta 500:21–69
Kurbanoglu S, Mayorga-Martinez CC, Medina-Sánchez M, Rivas L, Ozkan SA (2015) Antithyroid drug detection using an enzyme cascade blocking in a nanoparticle-based lab-on-a-chip system. Biosens Bioelectron 67:670–667
Kurokawa K, Mochizuki N, Ohba Y, Mizuno H, Miyawaki A, Matsuda M (2001) A pair of fluorescent resonance energy transfer-based probes for tyrosine phosphorylation of the CrkII adaptor protein in vivo. J Biol Chem 276:31305–31310
Lager I, Fehr M, Frommer WB, Lalonde S (2003a) Development of a fluorescent nanosensor for ribose. FEBS Lett 553:85–89
Lager I, Fehr M, Frommer WB, Lalonde S (2003b) Development of a fluorescent nanosensor for ribose. FEBS Lett 553:85–89
Lang HP, Hegner M, Gerber C (2005) Cantilever array sensors. Mater Today 8:30–36
Lee S, Myers WR, Grossman HL, Cho HM, Chemla YR, Clarke J (2002) Magnetic gradiometer based on a high-transition temperature superconducting quantum interference device for improved sensitivity of a biosensor. Appl Phys Lett 81:3094–3096
Lee YEK, Kopelman R, Smith R (2009) Nanoparticle PEBBLE sensors in live cells and in vivo. Ann Rev Anal Chem 2:57–76
Lim TC, Ramakrishna S (2006) A conceptual review of nanosensors. Z Naturforsch A 61:402–412

Love J (2007) Process automation handbook: a guide to theory and practice. Springer Science & Business Media, London

Lucas SW, Harding MM (2000) Detection of DNA via an ion channel switch biosensor. Anal Biochem 282:70–79

Mahato K, Prasad A, Maurya PK, Chandra P (2016) Nanobiosensors: next generation point-of-care biomedical devices for personalized diagnosis. J Anal Bioanal Tec 7:2

Mars A, Parolo C, de la Escosura-Muñiz A, Raouafi N, Merkoçi A (2016) Control of electron-transfer in immunonanosensors by using polyclonal and monoclonal antibodies. Electroanalysis 28:19

Medintz IL, Uyeda HT, Goldman ER, Mattoussi H (2005) Quantum dot bioconjugates for imaging, labelling and sensing. Nat Mater 4:35

Mizutani T, Kondo T, Darmanin S, Tsuda M, Tanaka S, Tobiume M, Asaka M, Ohba Y (2010) A novel FRET-based biosensor for the measurement of BCR-ABL activity and its response to drugs in living cells. Clin Cancer Res 16:3964–3975

Modi A, Koratkar N, Lass E, Wei B, Ajayan PM (2003) Miniaturized gas ionization sensors using carbon nanotubes. Nature 424:171–174

Mohsin M, Ahmad A (2014) Genetically-encoded nanosensor for quantitative monitoring of methionine in bacterial and yeast cells. Biosens Bioelectron 59:358–364

Mohsin M, Abdin MZ, Nischal L, Kardam H, Ahmad A (2013) Genetically encoded FRET-based nanosensor for in vivo measurement of leucine. Biosens Bioelectron 50:72–77

Mohsin M, Abidin MZ, Nischal L, Kardam H, Ahmad A (2016) Genetically encoded FRET-based nanosensor for in vivo measurement of leucine. Biosens Bioelectron 86:169–117

Monig CA, Jorgenson JW (1991) On-column sample grating for high-speed capillary zone electrophoresis. Anal Chem 63:802–807

Mousavi SM, Hashemi SA, Zarei M, Amani AM, Babapoor A (2018) Nanosensors for chemical and biological and medical applications. Med Chem 8:205

Mukhopadhyay R, Sumbayev VV, Lorentzen M, Kjems J, Andreasen PA, Besenbacher F (2005) Cantilever sensor for nanomechanical detection of specific protein conformations. Nano Lett 5:2385–2388

Nagai Y, Miyazaki M, Aoki R, Zama T, Inouye S, Hirose K, Iino M, Hagiwara M (2000) A fluorescent indicator for visualizing cAMP-induced phosphorylation in vivo. Nat Biotechnol 18:313–316

Nitin N, LaConte LE, Zurkiya O, Hu X, Bao G (2004) Functionalization and peptide-based delivery of magnetic nanoparticles as an intracellular MRI contrast agent. J Biol Inorg Chem 9:706–712

Noh HB, Lee KS, Chandra P, Won MS, Shim YB (2012) Application of a Cu–Co alloy dendrite on glucose and hydrogen peroxide sensors. Electrochim Acta **61**:36–43

Okumoto S, Looger LL, Micheva KD, Reimer RJ, Smith SJ, Frommer WB (2005) Detection of glutamate release from neurons by genetically encoded surface-displayed FRET nanosensors. Proc Natl Acad Sci U S A 102:8740–8745

Osterfeld SJ, Yu H, Gaster RS, Caramuta S, Xu L, Han SJ, Hall DA, Wilson RJ, Sun S, White RL, Davis RW, Pourmand N, Wang SX (2008) Multiplex protein assays based on real-time magnetic nanotag sensing. Proc Natl Acad Sci U S A 105:20637–20640

Paliwal A, Gaur R, Sharma A, Tomar M, Gupta V (2016) Sensitive optical biosensor on surface plasmon resonance using ZnO/au bilayered structure. Optik 127:7642–7647

Penza M, Antolini F, Antisari MV (2004) Carbon nanotubes as SAW chemical sensors materials. Sens Actuat B Chem 100:47–59

Perez JM, Josephson L, Weissleder R (2004) Use of magnetic nanoparticles as nanosensors to probe for molecular interactions. Chembiochem 5:261–264

Perez JM, Josephson L, O'Loughlin T, Högemann D, Weissleder R (2002) Magnetic relaxation switches capable of sensing molecular interactions. Nat Biotechnol 20:816–820

Perez JM, Simeone FJ, Saeki Y, Josephson L, Weissleder R (2003) Viral-induced self-assembly of magnetic nanoparticles allows the detection of viral particles in biological media. J Am Chem Soc 125:10192–10193

Peterson KE (1983) Silicon as a mechanical material. Microelectron Reliab 23:403
Rackus DG, Dryden MD, Lamanna J, Zaragoza A, Lam B (2015) A digital microfluidic device with integrated nanostructured microelectrodes for electrochemical immunoassays. Lab Chip 15:3776–3784
Sagadevan S, Periasamy M (2014) Recent trends in nanobiosensors and their application – a review. Rev Adv Mater Sci 36:62–69
Salonikidis PS, Zeug A, Kobe F, Ponimaskin E, Richter DW (2008) Quantitative measurement of cAMP concentration using an exchange protein directly activated by a cAMP-based FRET-sensor. Biophys J 95:5412–5423
San Martín A, Ceballo S, Ruminot I, Lerchundi R, Frommer WB, Barros LF (2013) A genetically encoded FRET lactate sensor and its use to detect the Warburg effect in single cancer cells. PLoS One 8:e57712
Schroeder V, Savagatrup S, He M, Lin S, Swager TM (2019) Carbon nanotube chemical sensors. Chem Rev 119:599–663
Sekretaryova AN, Eriksson M, Turner AP (2015) Bioelectrocatalytic systems for health applications. Biotechnol Adv 34:177–197
Shelby T, Banerjee T, Zegar I, Santra S (2017) Highly sensitive, engineered magnetic nanosensors to investigate the ambiguous activity of zika virus and binding receptors. Sci Rep 7:7377
Solin SA (2001) Semiconductor fridges get into shape. Phys World 14:29–30
Solin SA (2004) Magnetic field nanosensors. Sci Am 291:70–77
Song JM, Kasili PM, Griffin GD, Vo-Dinh T (2004) Detection of cytochrome C in a single cell using an optical nanobiosensor. Anal Chem 76:2591–2594
Shankaran DR, Gobi KV, Miura N (2007) Recent advancements in surface plasmon resonance immunosensors for detection of small molecules of biomedical, food and environmental interest. Sensors Actuators B Chem 121(1):158–177
Tan W, Shi ZY, Smith S, Birnbaum D, Kopelman R (1992) Submicrometer intracellular chemical optical fiber sensors. Science 258:778–781
Tang X, Bansaruntip S, Nakayama N, Yenilmez E, Chang YI, Wang Q (2006) Carbon nanotube DNA sensor and sensing mechanism. Nano Lett 6:1632–1636
Tu L, Klein T, Wang W, Feng Y, Wang Y, Wang JP (2013) Measurement of Brownian and Néel relaxation of magnetic nanoparticles by a mixing-frequency method. IEEE Trans Mag 49 (1):227–230
Turner AP (2013) Biosensors: sense and sensibility. Chem Soc Rev 42:3184–3196
Vallejo MF, Amo ML (2012) Optical fiber networks for remote fiber optic sensors. Sensors 12:3929–3951
Vinkenborg JL, Nicolson TJ, Bellomo EA, Koay MS, Rutter GA, Merkx M (2009) Genetically encoded FRET sensors to monitor intracellular Zn^{2+} homeostasis. Nat Methods 6:737
Vo-Dinh T, Griffin GD, Alarie JP, Cullum B, Sumpter B, Noid D (2000) Development of nanosensors and bioprobes. J Nanopart Res 2:17–27
Wang J, Lin W, Cao E, Xu X, Liang W, Zhang X (2017) Surface plasmon resonance sensors on raman and fluorescence spectroscopy. Sensors 17:2719
Wang SX, Li G (2008) Advances in giant magnetoresistance biosensors with magnetic nanoparticle. IEEE Trans Mag 44:1687–1702
Wu X, Zhao Z, Bai H, Fu T, Yang C, Hu X, Tan W (2015) DNA Aptamer selected against pancreatic ductal adenocarcinoma for in vivo imaging and clinical tissue recognition. Theranostics 5:985–994
Xu H, Aylott JW, Kopelman R, Miller TJ, Philbert MA (2001) A real-time ratiometric method for the determination of molecular oxygen inside living cells using sol– gel-based spherical optical nanosensors with applications to rat C6 glioma. Anal Chem 73:4124–4133
Yousefi AT, Tateno K, Bagher S, Tanaka H (2017) Development of frequency based taste receptors using bioinspired glucose nanobiosensor. Sci Rep 7:1623
Zhang J, Ma Y, Taylor SS, Tsien RY (2001) Genetically encoded reporters of protein kinase a activity reveal impact of substrate tethering. Proc Natl Acad Sci U S A 98:14997–15002

Zhang J, Qin Z, Zeng D, Xie C (2017) Metal-oxide-semiconductor based gas sensors: screening, preparation, and integration. Phys Chem Chem Phys 19:6313–6329

Zhang JX, Hoshino K (2018) Molecular sensors and Nanodevices: principles, designs and applications in biomedical engineering. Academic Press, London

Zhu JK (2001) Plant salt tolerance. Trends Plant Sci 6:66–71

Zuo L, Chen X, Yu S, Lu M (2016) Design and fabrication of a differential scanning nanocalorimeter. J Micromech Microeng 27:025006

Biosensor: An Approach Towards a Sustainable Environment

3

Neelam P. Negi and T. Choephel

Abstract

Of worldwide concern, ecological pollution affects human security and development. Presence of contaminants, particularly bacterial, viral, and parasitic pathogen poses serious health issues. Hence, there is an urgent requirement of developing techniques, which can rapidly recognize the pollutions for effective bioremediation processes. The source used for recognition, evaluation, and transformation of pollutants to non-pollutants to restore the ecological balance can be isolated enzymes or biological systems producing enzymes, as entire cells or in the immobilized state. For the detection and measurement of environmental pollution, biosensors are perfect and reliable. In this chapter, the present status of various kinds of biosensors and its applications are discussed.

Keywords

Biosensors · Transducers · Environmental monitoring · Bioelectronics

3.1 Introduction

In today's scenario, the major threat that we are facing by environmental pollution is rising day by day and affecting severe damage to the earth. The air pollution has increased with the modernization of the living standards and enhanced demands of consumers. This has in turn affected the increase of CO_2, other ozone-depleting substances, and dangerous particulate matter in the air and oil spills. Not only this, our surrounding consists of the soil which has also been affected, with the wide range of pollutants, viz., non-biodegradable materials, hazardous wastes, herbicides,

N. P. Negi (✉) · T. Choephel
University Institute of Biotechnology, Chandigarh University, Mohali, India
e-mail: neelam.biotech@cumail.in

M. Mohsin et al. (eds.), *Nanobiosensors for Agricultural, Medical and Environmental Applications*, https://doi.org/10.1007/978-981-15-8346-9_3

pesticides, and different chemicals. Various compounds that are being used by mankind are directly linked with the one or other type of pollution (Claude et al. 2007). These chemicals are very important application wise and are required by all of us in the modern era of lifestyle, e.g. household care products, pesticides, medicines, cosmetics, etc. (Nigam and Shukla 2015). Apart from these man-made pollutants, biological pollutants such as viruses and bacteria are also contaminating the water resources which have raised serious concerns regarding their pathogenicity potential as majority of micropollutants are cause for serious fatal diseases (Wang et al. 2014).

Detailed analysis of toxicants and its regular evaluation are of utmost importance, and it should be comprehensive so as to provide safety and security to mankind. In the present era of environmental friendly sciences, biological agents are employed to do the analysis of toxicants in the air, water, and soil. These include bacteria, fungi, blue green algae, and enzymes, etc. Among these, bacteria are good options because of their unicellular nature, rapid growth, enough population size, low cost of culturing and maintenance, and also these can be genetically modified to analyse the pollutants in their environmental conditions. For the overall biological system security and environmental safety, proper recognition and observation of environmental pollutants are important. Conventional tools for analysis such as spectroscopy and chromatography are skill demanding, consume more time in analysis, and expensive. Therefore there is an urgent need of a quick, potable, easy, and specific method for exploring environmental security threats. The necessity of devices for environmental monitoring has encouraged the development of more suitable approaches and new advanced tools to observe the analytes of environmental significance as cheap, reliable, and fast (Lang et al. 2016; Hassani et al. 2017). Therefore, a basic, explicit, delicate, fast, and versatile strategy should be produced for the investigation and perception of ecological security dangers, and subsequently, biosensors show up as a suitable decision as a systematic apparatus for such examination (Podola et al. 2004).

A biosensor is an integrated device that composed of two components, i.e. bioreceptors and transducer. Bioreceptor produces a physicochemical reaction in response to an analyte, whereas the transducer converts these physicochemical signals into an electrical signal, which is then converted into an electrical display. These sensors have high specificity and sensitivity that provide continuous real-time signals. It is used in the investigation of biomaterial samples, i.e. quantity, composition, structure, and function, for various applications including fermentation process control, food and environmental quality control, agriculture, military and diagnostic sectors. One more type of biosensors, i.e. phage biosensor has been developed to screen pathogens and contaminants in sustenance and condition (Van Dorst et al. 2010). The significant toxin that can be progressively recognized and evacuated utilizing biosensors incorporates substantial metals, toxins, pesticides, heavy metals, biochemical oxygen demand (BOD), nitrogenous mixes, and different pathogens (counting numerous infections and microorganisms).

3.2 Classification of Biosensor

A biosensor is an independent essential apparatus that gives exact, quantitative, and systematic data, utilizing a biochemical receptor that is in direct contact with a transduction component. The biosensor is principally comprised of three constituents: biological component, a transducer, and a signal processing unit (Turner 2013). An illustration of a biosensor is given in Fig. 3.1.

A biosensor can be categorized on the basis of biological element; it could be (enzyme, antibody, aptamer) and on the basis of transducers it could be (electrochemical, optical, piezoelectric, electrical, and colorimetric) (Fig. 3.2).

3.2.1 Biological Element

3.2.1.1 Enzyme-Based Biosensor

Enzyme performing as biosensors characteristically catalysed redox reaction. It can be observed by various electrochemical methods. To know the physiological impact of pesticides in the food safety environment and quality control some enzyme-based biosensors (Verma and Bhardwaj 2015) were established. Acetylcholinesterase (AChE) inhibition-based biosensors were established for this persistence (Fournier 2005).

3.2.1.2 Antibody-Based Biosensor

Antibodies have been considered as best recognition motifs in biosensors and it is made up of individual amino acids. The use of antibodies as recognition segments has dragged in contemplations especially with the approach of monoclonal antibody production. Antigen–antibody sensing is highly sensitive for the detection of the immunogen. Entrapment of antibodies on the surface of biosensors is very much important for increasing the efficiency of the system. The reproducibility, sensitivity,

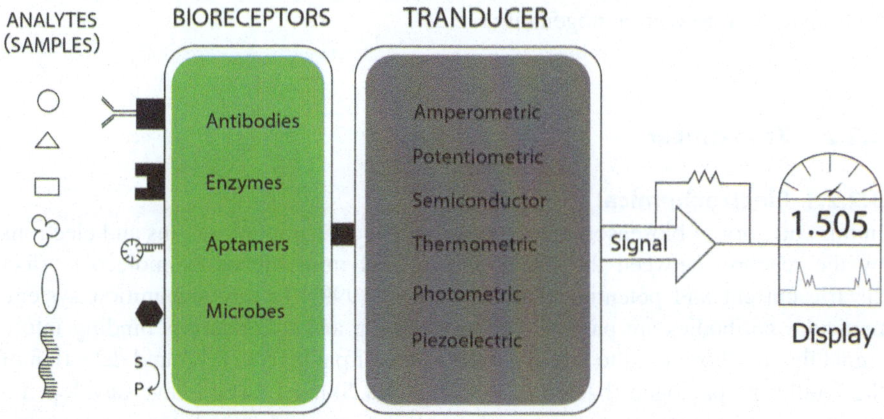

Fig. 3.1 Segments of a biosensor

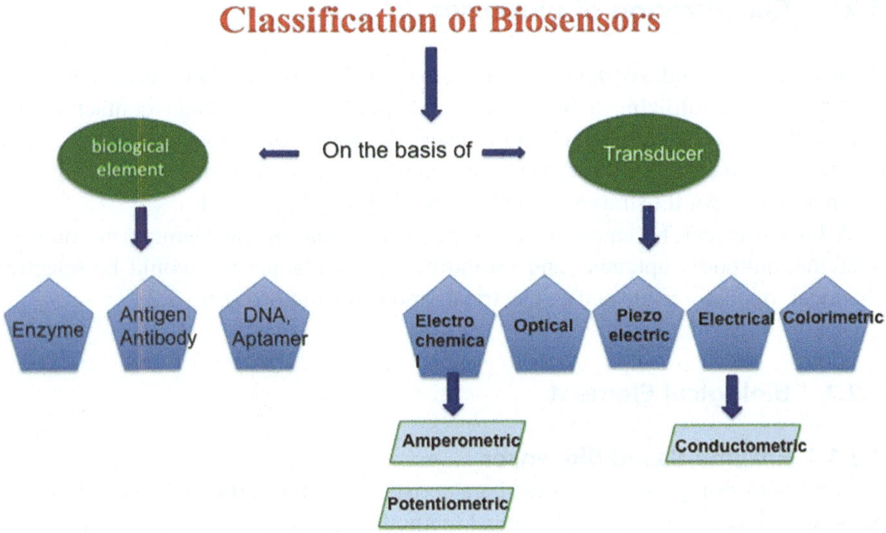

Fig. 3.2 Classification of biosensors

and selectivity of the system solely depend upon the specific interactions between antigen and antibody (Schramm et al. 1993). Thusly the enzymatic based sensors which include glucose oxidase leads to the oxidation of glucose into gluconic acid.

3.2.1.3 Aptamer Based Biosensor

Aptamers are DNA or RNA that can be synthesized beside an inclusive variety of target analytes having strong binding affinities (Wang et al. 2015). The development and production of antibodies depend on the immunization of mammals through cell culture. Aptamers generated by in vitro techniques without cells are easy and economical. Therefore aptamers can be chosen against any type of target either high toxicity or low immunogenicity.

3.2.2 Transducer

3.2.2.1 Electrochemical

In this category of biosensors the production and consumption of ions and electrons by the reaction between the target analyte and immobilized biomolecule affect electric current and potential (Thévenot et al. 1999). In this recognition system, particular antibodies are paired with an electrode and it converts a binding into a signal that has been used to detect a wide range of pathogens. For rapid detection of the foodborne pathogen (bacteria), Pohanka and Skládal (2008) have developed a functionalized biosensor that is based on electrochemical impedance spectroscopy (EIS).

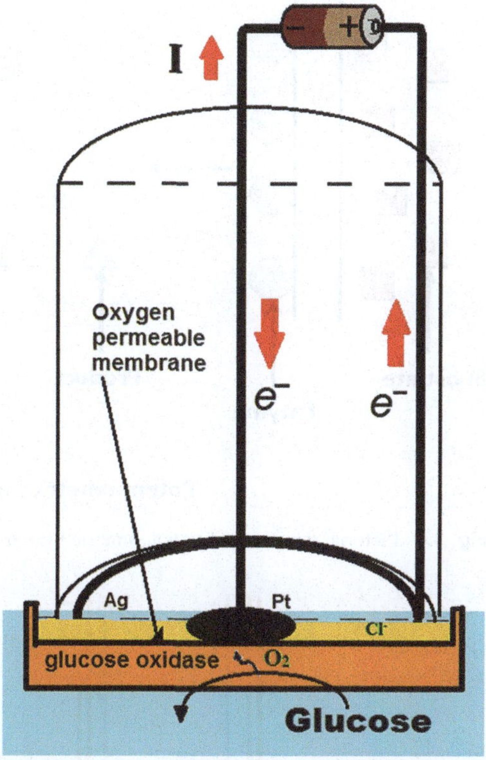

Fig. 3.3 Diagrammatic depiction of amperometric biosensors

3.2.2.2 Amperometric

Amperometric biosensors measure change in current and are based on the movement of electrons that is current determination resulting from enzyme-catalysed redox reaction (Okon and Ronkainen 2017) (Fig. 3.3).

3.2.2.3 Potentiometric

Potentiometric biosensor measures the potential difference between the electrodes when no current flow. pH electrodes are mostly used in this type of biosensor and the changes in the ionic concentration are determined by the ion-selective electrodes. Other important electrodes are ammonia selective and carbon dioxide selective electrode. These inferences remain appropriate to environmental, analysis of food and medical field. By the use of immobilized cells of *S. ellipsoids* a potentiometric oxygen electrode was well retained for the measurement of the ethanol (Rogers and Gerlach 1996). The potentiometric microbial biosensor is built utilizing a particle specific anode encompassed with the immobilized microbial cells on it. (Fig. 3.4) (Mulchandani et al. 1998).

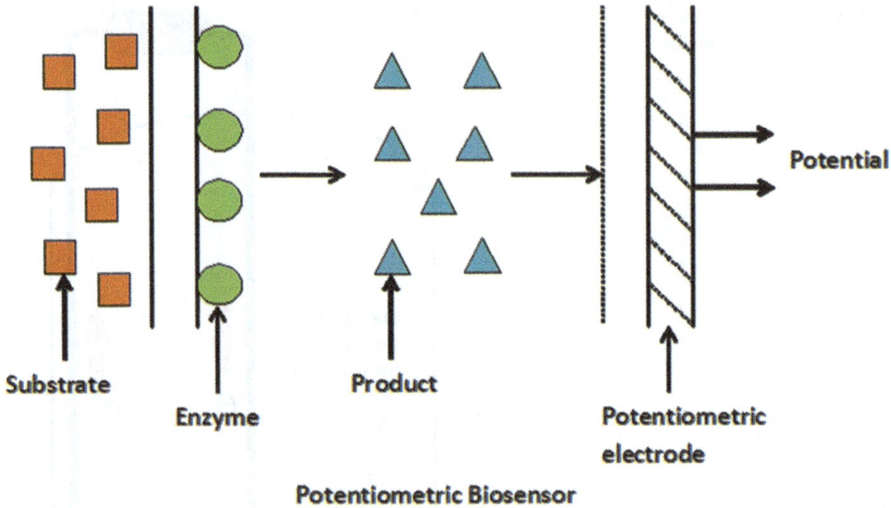

Fig. 3.4 Pictorial depiction of potentiometric biosensors

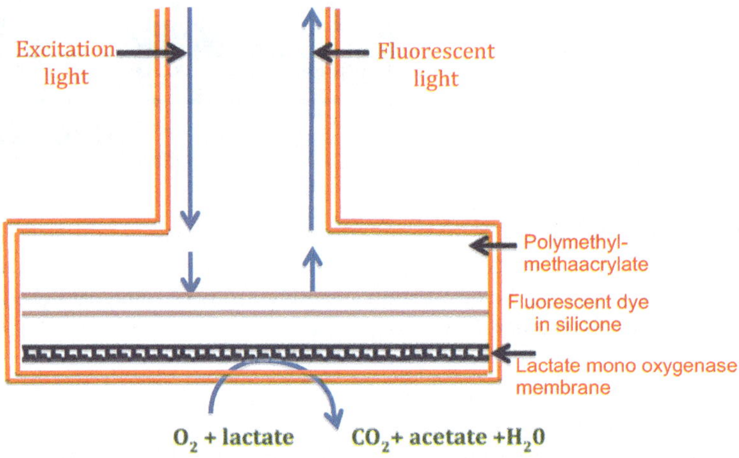

Fig. 3.5 Pictorial depiction of fibre optic biosensors

3.2.2.4 Optical

Optical biosensors are used for the detection of pollutants in the environment, where the reaction of the enzyme with the specific analyte is monitored (Fig. 3.5) (Rodriguez-Mozaz et al. 2004). In this case, total reflection phenomena of light is used to travel through fibre optic and it produces efficient wave at the border of the fibre optic which provide energy for the surface to get excited and the fluorescence emitted by the molecules at the surface is captured by optical fibre. In real-time, process-monitoring bioluminescence uses the arrival of illumination by specific microorganisms and the gene responsible for the reflection of light via total internal refection phenomena, the bacterial structure was identified by the fluorescence. A few optical biosensors are made for estimating Hg^{2+} by consolidating the regulatory part of the mer operon gene (merR) and lux CDABE of the bacterial system. The binding of Hg^{2+} activates Mer promoter and merR enacts, which leads to the synthesis of the reporter gene (lux) pursued by light generation (Shukla et al. 2013). Testing the copper content in the soil is checked by optical biosensors utilizing recombinant *Pseudomonas fluorescens* (Reyes De Corcuera and Cavalieri 2003). In optical immunosensor, antigen and antibodies ligand interaction generates a detectable optical signal.

3.2.2.5 Piezoelectric

Piezoelectric biosensors are based on the pairing of the bioelement with a piezoelectric component, commonly used material is quartz coated by gold electrodes. Other materials are aluminium nitride, tantalite, zinc oxide, etc. It is used in sound vibrations. These transducers permit the tag-free identification of molecules. By the application of a signal of a specific frequency, these quartzes can be made to oscillate. Frequency of vibration is depending on the electrical frequency that is applied to the crystal together with the crystal's mass. These biosensors displayed potential applications in the environment, food and medical field (Tombelli et al. 2005) (Fig. 3.6).

3.2.2.6 Electrical

Conductometric

There are some reactions in the biological classification that change the ion concentration and mobility with the help of these ionic species. Thus, the electrical

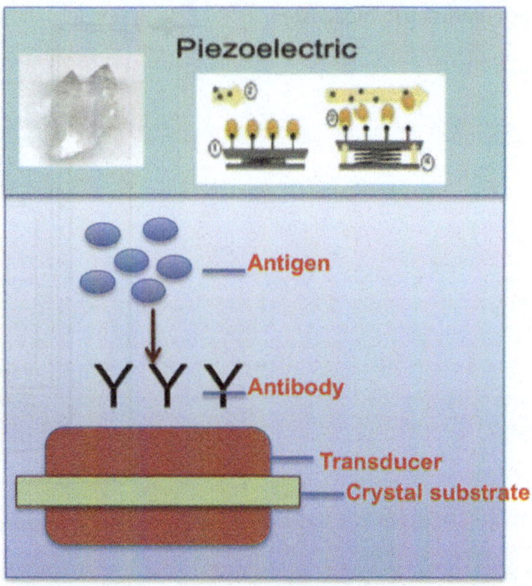

Fig. 3.6 Pictorial depiction of piezoelectric biosensors

conductivity can be measured (Soldatkin et al. 2014). Urea biosensor is the example of conductometric biosensor applying immobilized urease (Velychko et al. 2016) catalysing the following reaction:

$$\text{Urea} + 2H_2O + H^+ \rightarrow 2NH_4^+ + HCO_3^-$$

Colorimetric

Some organic responses in the biological structure are related to measure the temperature change during a biochemical reaction. This method is used for diagnosing organic as well as inorganic species (Pisoschi 2016). They are more normally indicated to as heat biosensors or colorimetric biosensors (Fig. 3.7).

3.3 Biosensors for Environmental Monitoring

As a result of technological development, huge utilization and production of chemicals and by-products industries have been released in the environment. Toxins, heavy metals, and pesticides are contaminants known to influence environmental quality. As an outcome, a variety of particular biosensors have been grown extensively and applied to their environmental determination. For pesticides like organophosphorus and carbamate, different enzymatic biosensors dependent upon the action of the choline oxidase and on the hindrance of acetylcholinesterase (AChE) and butyrylcholinesterase (BChE) have been created (Andreou and Clonis 2002). On another hand, the DNA sensor is a sort of contamination pointer of environmental samples and monitors the collaboration of little pollutants with the immobilized DNA layer (Diculescu et al. 2016).

Fig. 3.7 Pictorial depiction of colorimetric biosensors

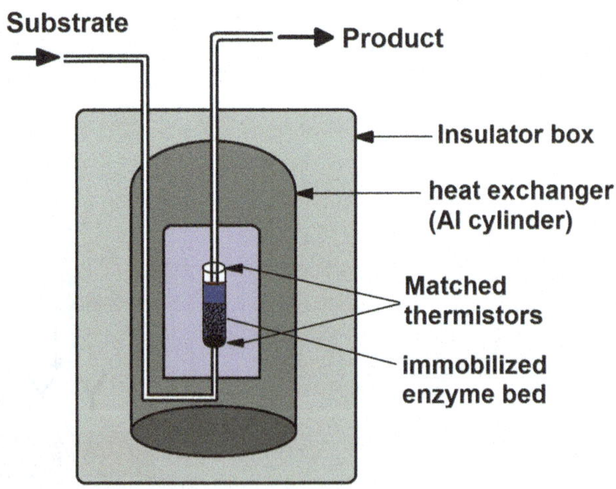

3.3.1 Toxin

For the detection of contamination such as heavy metals in an environmental sample, most of the bacterial biosensors used are genes, which are resistant against these metals. Some of the bacterial systems have been assessed for the detection of copper, zinc, mercury, cobalt, etc. Metal ion detection can be done by the enzyme-catalysed reaction because the majority of the ions hinder specifically at less enzyme concentration. By using the genetic engineering techniques scientists have modified the sites of allosteric enzymatic and build up the recombinant proteins to drive the dependability, stability, selectivity, and affectability of measurements. Acetylcholinesterase (AChEs) separated from electric eel is generally utilized in the bioassay and biosensor for the identification of insecticides but it lacks sensitivity against many insecticides. Hence Fournier (2005) produced numerous mutants of AChEs to develop more selectivity towards many insecticides such as carbamates, organophosphate, and natural neurotoxin. For the identification of AChE inhibitors, the AChE-immobilized transducer surfaces have been used.

Another way is site-directed mutagenesis to deal with upgrade bioassay affectability regarding analysis of insecticides and toxins. The hereditary adjustment of the catalyst has been appeared to yield increasingly dynamic and stable bioreceptor particles (Schulze et al. 2003). The recombinant enzyme was utilized for the identification of aflatoxin B1 (AFB1) to improve the stability and sensitivity of bioassay (Arduini et al. 2007).

3.3.1.1 Recombinant Fragments of Antibody

Immunoglobulin (IgG) is ideal in the improvement of immunosensors such as natural toxins such as mycotoxins like Deoxynivalenol (DON), fumonisin B1 (Pagkali et al. 2018) and zearalenone (Foubert et al. 2019), phytotoxin and cyanotoxins (Cunha et al. 2018). Phage display advancement has made ease to determine the recombinant antibodies from the large antibody libraries against toxins. These libraries are helpful for the screening of bacterial environmental toxins like Shiga, fumonisin B1, deoxynivalenol, and zearalenone, (Wang et al. 2012a, b; Pagkali et al. 2018 and Foubert et al. 2019). A specific sequenced scFv helps to test the toxin which is acquired from the monoclonal antibody from mycotoxin like fumonisin B_1 where its binding character possesses 12 times lesser than the Mabs of LOD > 0.11 g/kg (Min et al. 2009).

3.3.1.2 DNAzymes

It has the capacity with respect to the advancement of biosensors for a wide scope of toxins. DNAzyme makes fluorescent biosensors with the fluorophore and quencher in the recognition medium and DNA cleavage creates a flag (Yu et al. 2013). The coupling of gold nanoparticles (AuNPs) with DNAzyme was investigated to develop colorimetric biosensors. The particular scheme permitted DNAzyme to catalyse the H_20_2 in the presence of hemin as peroxidase that mimics to produce chemiluminescence response for the detection of toxins (Xiao et al. 2004).

Table 3.1 Biosensors for the toxic compounds detection

Analytes	Biosensing elements	Transducers	References
(Heavy metals)			
Mercury, cadmium, arsenic	Urease enzyme	Electrochemical	Pal et al. (2009)
Mercury(II) and lead (II) ions	DNA	Optical	Knecht and Sethi (2009)
Cadmium, copper, and lead	Sol–gel-immobilized urease	Electrochemical	Ilangovan et al. (2006)
(Phenolic compounds)			
Phenol	Mushroom tissue (tyrosinase)	Amperometric	Silva et al. (2010)
Phenol, p-cresol, m-cresol and catechol	Polyphenol oxidase	Amperometric	Karim and Fakhruddin (2012)
m-cresol or catechol	DNA	Amperometric	Claude et al. (2007)
(Pesticides)			
Parathion	Parathion hydrolase (biocatalytic)	Amperometric	Mostafa (2010)
Carbaryl	Acetylcholinesterase	Amperometric	Mostafa (2010)
Paraoxon	Alkaline, phosphatase	Optical	Mostafa (2010); Sassolas et al. (2012)
(Herbicides)			
2,4-Dichloro-phenoxy acetic acid	Acetylcholinesterase	Amperometric	Sassolas et al. (2012)
Diuron, paraquat	Cyanobacteria	Bioluminescence	Sassolas et al. (2012)

3.3.1.3 MIPs

Molecularly imprinted polymers (MIPs) were portrayed for recognition of mycotoxins and other minute compounds. MIPs have been employed in immuno-based sample processing strategies which have given an account for toxin detection (Xu et al. 2011) (Table 3.1).

3.3.2 Pesticides

Pesticides have been used extensively to protect the crops by diminishing pests and weeds for the benefit of the agriculture sector. It is expected that the utilization of pesticides support to save yield of approx. 30% around the globe (Nougadère et al. 2011). Pesticides are dangerous mixtures utilized against harmful pests and get into the food industry by means of bringing about health issues by extreme exposure to a harmful chemical residue in pesticides (Mostafalou and Abdollahi 2013). Because of their stability and harmfulness it is quick to identify the pesticide. Recently, nanotechnology and nanomaterial have been considerably assigned as an integrated

Table 3.2 Analytical characteristics of OPs and its biosensors

OPs	Sensing mechanism/ Transduction	References
Malathion	Electrochemical (amperometry)	Gahlaut et al. (2012)
	Immunosensor (amperometric)	Prabhakar et al. (2007)
Phosalone	Chemiluminescence	Zhi-tao et al. (2010)
Chlorpyrifos	Immunosensor (amperometric)	Prabhakar et al. (2007)
	Label-free immunosensors SPR	Mauriz et al. (2007)
Dimethoate	Electrochemical (amperometric)	Hui-li (2005)
Chlorfenvinphos	Quartz crystal microbalance (piezoelectric)	Halamek et al. (2005)
Phorate, profenofos, isocarbophos, omethoate	Fluorometric FRET, aptamer-type DNA	Wang et al. (2012a, 2012b); Zhang et al. (2014)

section of biosensor technology for the detection of pesticides. These techniques developed with small size, low cost, and simple design which shown adequate outcomes for environmental health, food quality analysis, and toxicity detection (Guler et al. 2010). Innovative trends on pesticide assessment have been introduced an aptamer which is an alternative to the standard antibodies and provides more selective biosensors for pesticide detection. Also, molecularly imprinted polymers (MIPs) are considered as novel recognition materials for the detection of environmental pollution (Sassolas et al. 2012). Table 3.2 shows the organophosphorous (OPs) examples with their relevance biological effects.

3.3.3 Pathogen

Bacteria such as *Staphylococcus aureus*, *E.coli*, and *Pseudomonas aeruginosa* are the distinct source of nosocomial diseases in case of hospital. To lessen their effect on health, researchers should recognize them proficiently.

3.3.3.1 Antimicrobial Peptides

Antimicrobial peptides (AMPs) are short sequence peptides, with usually fewer than 50 amino acid residues reported in living systems. Antibiotics target for particular proteins, different AMPs against bacterial membranes. In the area of pathogen identification, these antimicrobial peptides have been utilized (Dao et al. 2018). AMPs could be proficient as bioreceptors for biosensing applications. Soares et al. (2004) were among the first to utilize the biosensor to explore the capability of antimicrobial peptides and also investigated the coupling capacities of AMPs (Cecropin A, Cecropin P1, Pleurocidin, PGQ, and SMAP-29) that were synthesized chemically and adjusted by adding C-terminal cysteine to immobilize onto maleimide responsive plates. These had the capacity to tie *E. coli* O157: H7 especially than *S. aureus* 27217 (Soares et al. 2004).

3.3.3.2 Sugars/Lectins

Bacteria can be recognized with by its lectin/starch. The Quartz Crystal Microbalance (QCM) detecting layer was developed as for settling of Concanavalin A over the mannose-monolayer and a couple of hundred E. coli W1485/ml was determined. Phenyl and aliphatic-adjusted sugar ligands were studied and low recognition levels for various microscopic organisms were determined with the help of surface plasmon resonance (electrochemical identification forms) (Yazgan 2014).

3.3.4 Heavy Metals

Heavy metals are the reason for serious pollution issues in present time. As a result of non-biodegradable metal, even in a little concentration, they are a threat to the animal and surrounding (Peavy et al. 1988). Hence, variety of enzymes has been employed for the heavy metal analysis and it relies upon activation or reduction of their significant properties. The situation of the metal inhibition of enzymes depends upon the reaction of metal with thiol or methyl thiol groups of amino acids (Corbisier et al. 1999). The reduction of chromium level has been indicated utilizing lactate dehydrogenase, pyruvate kinase, and hexokinase. Hence, it helps to sort out different pollutants such as metal ions and they are based on high technological models (Cowell et al. 1995). For the detection of metals like Ag, Cu, Hg, Pb, and Zn, a responsive or sensitized biosensor was used where L-lactate oxidase integrated with L-lactate dehydrogenase coating onto the O_2 electrode was constructed with the condition of lactate dehydrogenase inhibition. Along with EDTA, KCN, and DTT, the sensor was regenerated within 10 mins (Fennouh et al. 1998). Mercury inhibits the peroxide reductase once the immobilization on the polystyrene plate, in chitosan film, even on the surface of chromatography paper (Shekhovtsova et al. 1997). For determination of mercury, an optical fibre biosensor depends upon the urease immobilized on controlled pore glass has been initiated. Such a biosensor gives a respond within 12 mins (Andres and Narayanaswamy 1995). Hg, Cu, Cd, and Zn inhibition of urease was analysed applying the urease glutamic dehydrogenase system as the biological component integrated with amperometric transduction (Rodriguez et al. 2004).

3.3.4.1 Protein-Based Biosensor

A diversity of non-enzymatic proteins or modified proteins that are developed to link a particular metal for biosensor development. An integration of synechococcal SmtA metallothionein with glutathione-S-transferase approach was also used in a biosensor. Engineered cyanobacterial protein was immobilized on 1-Ethyl-3-(3-dimethylaminopropyl) carbodiimide, resulted in higher sensitivity and broad selectivity towards heavy metals like Hg, Zn, Cd, and Cu (i.e. fM levels) (Corbisier et al. 1999).

3.3.4.2 Antibody-Based Biosensor

Developed immunoassay as a possible outcome for determination of metal ions with high selectivity, specificity, and sensitivity is applicable to the precise waste particles for which specific antibodies are produced (Blake, 1995). More recently, monoclonal antibodies for the composite of EDTA with other metals, cobalt-diethylenetriaminepentaacetic acid (Co-DTPA), and lead-cyclohexyl diethylenetriamine-pentaacetic acid (Pb-CHXDTPA) with sensitive are developed (Blake et al. 2001) and there are also other devices used for detection of pollutants such as microwell ELISA format.

3.3.4.3 Genetically Engineered Microorganism (GEM)-Based Biosensors

Microbes, which are highly resistant to heavy metal, are specific to a few metals and a transformed bacterium for assessing metal like Cd and Pb has been succeeded. A sensor (pT 0024) included as a plasmid sensor that transports luciferase reporter gene which is present under the restrict of *Staphylococcus aureus* cadA promoter shows expression in *Bacillus subtilis* and *Staphylococcus aureus* strain. A response was reported in *Staphylococcus aureus strain* for Cd and Pb with detection limits of 10 nM and 33 nM (Tauriainen et al. 1998). Likewise, the soil bacterium *Ralstonia eutropha* and currently as *R. metallidurans* is employed for detecting available metals by using promoter cassettes-lux CDABE monitoring system (Corbisier et al. 1999). The possibility of upcoming microbic biosensor depends on genetically modified microbes for heavy metals and toxic chemical (D'Souza 2001). Other metals and transducers are given in Tables 3.1 and 3.3.

Table 3.3 Biosensors for heavy metals detection

	Heavy metal	Devices	References
Enzyme			
Peroxidase	Hg	Colorimeter (oxidation-based)	Shekhovtsova et al. (1997)
Urease	Hg	Fibre optic	Andres and Narayanaswamy (1995)
Urease	Hg Cu Cd Zn	Amperometer	Rodriguez et al. (2004)
Antibody			
Monoclonal antibodies	Cd Co Pb	KinExATM immunoassay	Blake et al. (2001)
Microbes			
Staphylococcus aureus^{+}	Cd Pb	Luminometer	Tauriainen et al. (1998)
Escherichia coli+	Cu Cr Pb	Amperometer	Corbisier et al. (1999)

3.3.5 Biological Oxygen Demand (BOD)

BOD represents the amount of the dissolved oxygen necessary for the aerobic biological organism to breakdown organic compound present in a water sample. It could be a major property for the effectualness assessing treatment. To reduce environmental pollution, unleash of cytotoxic compounds and health issues, water quality assessment is important and mandatory (Ruban et al. 2001). Biochemical oxygen demand measures the requirement of dissolved oxygen for the aerobic organism. Biochemical oxygen demand (BOD) sensors allow monitoring with accuracy, high speed, and high stability. Microbial fuel cells (MFCs) are the developing technologies wherever the electric currents generated correspond to the concentration of the biodegradable compound, took up at the anoxic anode. Microbial fuel cells display the ability for the development of feasible BOD sensors with low maintenance. Moreover, the account of coulometric sensors that have a distinct background of detection to amperometric sensors (the measurement of the current with respect to BOD). According to (Modin and Wilén 2012) an acetic acid-fed microbial electrolysis cell (MEC) was used as a colorimetric sensing element. An increase in the concentration of substrate enables saturation of the electrode-biofilm and the level of current gives a further rise in the concentration yielding no alteration in the current (Kim et al. 2013). This prevents the application of an MFC primarily based detector for detecting water pollutants from various water sources. MFCs (serpentines, tubular, or multi-staged) which are connected hydraulically have a special feature because from the primary MFC, substrate flows to the cells (electrodes) where anodic biofilms acclimatized to low substrate concentration consumes remaining substrates (Chung and Okabe 2009). Nonetheless, this improves the potential for improving the range of biosensor.

3.4 Biosensors and Bioelectronics on a Smartphone

3.4.1 Microscopic Bioimaging on a Smartphone

Biosensor and bioelectronic devices has achieved success in various fields like transmission of drugs, clinical diagnosis, food and environment monitoring. They produced substantial tools for analysis and detection as standard physical and chemical sensing elements. Microfabricated transducers and small devices are used to acquire real-time, point of care testing, and detections for various range of analytes easily, particularly for the medical and environmental trials (Sang et al. 2015).

With the nanotechnology, biosensors are often minimized to small scale and merged into chip devices as the detectable arrays (Ferrier et al. 2015). Bioelectronic (biosensor) on the smartphone was already developed from biosensing attempts utilizing old cell phones referred to as featured phone (Table 3.4).

Table 3.4 For biochemical detection smart phone-based optical biosensors

Techniques	Sensing mechanism	Recognition element	Analytes	References
Fluorescence	Fluorescent microscopy	Alexa-488 label	HCMV	Wei et al. (2013)
	Fluorophore quencher	Molecular beacon probe	MicroRNA-21	Yu et al. (2014)
	Fluorescent microscopy	Anti-*E. Coli*/ *Salmonella*	*E. Coli*/ *Salmonella*	Nicolini et al. (2015)
Colorimetric	ELISA	Anti-BDE-47	BDE-47	Chen et al. (2014)
	Chemiluminescent	Glucose oxidase	Glucose	Chun et al. (2014)
	Chemiluminescent/ LFIA	Anti-cortisol	Salivary cortisol	Zangheri et al. (2015)
	Mie scattering	Reagentless	E. Coli (on beef)	Liang et al. (2014)
Imaging	Shadow imaging microscopy		Blood cell/ microorganism	Lee and Yang (2014)
	Brightfield imaging	Anti-A/B/D antibodies	Blood typing	Guan et al. (2014)
	Brightfield microscopy		*P. falciparum*/ sickle cell	Breslauer et al. (2009)

3.5 Future Perspective

The advancement of biochemistry, chemistry, physics, engineering, electronics, and bioelectronics will distinctly impact the subsequent biosensor production. Biosensors technology succeeds on the basis of transducer and sensing technology development. The cost of sophisticated instrumentation is costly but technology progression leads to decrease in component costs. However, since biosensor sensitivity and selectivity depend on the properties of the biorecognition elements, a pivotal site of the future biosensor is the improvement of molecular recognition elements. For biosensor applications, genetic engineering relies on genetically engineered receptor molecules and genetically transformed cells. Recombinant antibodies are produced after the design of the amino acid sequence of the antibodies binding region. In addition, phage display libraries can be applied to select, isolate, and analyse the specific antibodies or peptides with binding affinities to monoclonal antibodies for any suitable antigen (Iqbal et al. 2000). Scientists are now investigating the potential to use phage-displayed peptides as reagents for biosensor applications. (Hock et al. 2002) Biosensors other than enzyme-based should be developed in the future so that the sensitivity and overall performance level of the system can be accentuated. Aptamers are novel recognition elements able to bind similarly to antibodies to a target molecule with specificity and selectivity. They can be used in aptasensors as well as aptazymes (Luzi et al. 2003). On the other hand, modified cell biosensors are also being designed, such as inducing bioluminescence to a modified gene with the target component, such recombinant bacterial sensors are developed by for the determination of Cr, Cd, Hg, and Zn. The environment applications of biosensor technology are not investigated fully as compared to medical applications. Large-scale biosensor technology is associated with the medical field and some are personalized for the examination of environmental samples. Hence huge effort is needed for the development of consistent devices used for pesticide detection. Together with these technologies, we need to focus on the stability, sensitivity, and selectivity of the biocomponent.

3.6 Conclusion

Over the past period, fast development occurs in the field of biosensors in the research level as well as the development of the product. A wide range of techniques can be used for the development of biosensors with a high-affinity reagent to the specific analytes for environmental analysis. The biosensor devices are sensitive, flexible, easy to use, and selective, hence they are perfect tools for monitoring of environmental applications. Contrast with biomedical applications, environmental applications of biosensor advances are still in their earliest stages and face different difficulties because of the inherent qualities of natural examination. Different nanomaterials are utilized in the advancement of biosensors for ecological methodology and investigation. Although biosensors are developed for the detection of environment analytes, very few biosensors are commercially available. With high throughput analysis of samples, huge efforts are in need towards innovations on the development of biosensors. Hence, pollution can be controlled easily and diminish environmental issues.

References

Andreou VG, Clonis YD (2002) A portable fibre-optic pesticide biosensor based on immobilized cholinesterase and sol-gel entrapped bromocresol purple for in-field use. Biosens Bioelectron 17 (1–2):61–69

Andres RT, Narayanaswamy R (1995) Effect of the coupling reagent on the metal inhibition of immobilized urease in an optical biosensor. Analyst 120:1549–1554

Arduini F, Errico I, Amine A, Micheli L, Palleschi G, Moscone D (2007) Enzymatic spectrophotometric method for Aflatoxin B detection based on acetylcholinesterase inhibition. Anal Chem 79 (9):3409–3415

Blake DA, Jones RM, Blake RC, Pavlov AR, Darwish IA, Yu H (2001) Antibody-based sensors for heavy metal ions. Biosens Bioelectron 16(9–12):799–809

Breslauer DN, Maamari RN, Switz NA, Lam WA, Fletcher DA (2009) Mobile phone based clinical microscopy for Global Health applications. PLoS One 4(7):e6320

Chen A, Wang R, Bever CRS, Xing S, Hammock BD, Pan T (2014) Smartphone-interfaced lab-on-a-chip devices for field-deployable enzyme-linked immunosorbent assay. Biomicrofluidics 8 (6):064101

Chun HJ, Park YM, Han YD, Jang YH, Yoon HC (2014) Paper-based glucose biosensing system utilizing a smartphone as a signal reader. Bio Chip J 8(3):218–226

Chung K, Okabe S (2009) Continuous power generation and microbial community structure of the anode biofilms in a three-stage microbial fuel cell system. Appl Microbiol Biotechnol 83:965–977

Claude D, Houssemeddine G, Andriy B, Jean-Marc C (2007) Whole cell algal biosensors for urban waters monitoring. Ther Nova 7:1507–1514

Corbisier P, Lelie D, Borremans B, Provost A, Lorenzo V, Brown NL, Lloyd JR, Hobman JL, Cso¨regi E, Johansson G, Mattiasson B (1999) Whole cell-and protein-based biosensors for the detection of bioavailable heavy metals in environmental samples. Anal Chim Acta 387:235–244

Cowell DC, Dowman AA, Ashcroft T (1995) The detection and identification of metal and organic pollutants in potable water using enzyme assays suitable for sensor development. Biosens Bioelectron 10(6–7):509–516

Cunha I, Biltes R, Sales M, Vasconcelos V (2018) Aptamer-based biosensors to detect aquatic phycotoxins and cyanotoxins. Sensors 18(7):2367

Dao TNT, Lee EY, Koo B, Jin CE, Lee TY, Shin Y (2018) A microfluidic enrichment platform with a recombinase polymerase amplification sensor for pathogen diagnosis. Anal Biochem 544:87–92

Diculescu VC, Chiorcea-Paquim AM, Oliveira-Brett AM (2016) Applications of a DNA-electrochemical biosensor. TrAC Trends in Anal Biochem 79:23–36

D'Souza SF (2001) Microbial biosensors. Biosens Bioelectron 16:337–353

Foubert A, Beloglazova NV, Hedström M, De Saeger S (2019) Antibody immobilization strategy for the development of a capacitive immunosensor detecting zearalenone. Talanta 191:202–208

Fennouh S, Casimiri V, Meyer AG, Burstein C (1998) Kinetic study of heavy metal salt effects on the activity of L-lactate dehydrogenase in solution or immobilized on an oxygen electrode. Biosens Bioelectron 13:903–909

Ferrier D, Shaver M, Hands P (2015) Micro- and nano-structure based oligonucleotide sensors. Biosens Bioelectron 68:798–810

Fournier D (2005) Mutations of acetylcholinesterase which confer insecticide resistance in insect populations. Chem Biol Interact 157-158:257–261

Gahlaut A, Gothwal A, Chhillar AK, Hooda V (2012) Electrochemical biosensors for determination of organophosphorus compounds: review. Open J Appl Biosens 1(01):1

Guan L, Tian J, Cao R, Li M, Cai Z, Shen W (2014) Barcode-like paper sensor for smartphone diagnostics: an application of blood typing. Anal Biochem 86(22):11362–11367

Guler G, Cakmak Y, Dagli Z, Aktumsek Aand Ozparlak H (2010) Organochlorine pesticide residues in wheat from Konya region, Turkey. Food Chem Toxicol 48(5):1218–1221

Halamek J, Prˇibyl J, Makower A, Skládal P, Scheller F (2005) Sensitive detection of organophosphates in river water by means of a piezoelectric biosensor. Anal Bioanal Chem 382(8):1904–1911

Hassani S, Momtaz S, Vakhshiteh F, Maghsoudi AS, Ganjali MR, Norouzi P, Abdollahi M (2017) Biosensors and their applications in detection of organophosphorus pesticides in the environment. Arch Toxicol 91:109–130

Hock B, Seifert M, Kramer K (2002) Engineering receptors and antibodies for biosensors. Biosens Bioelectron 17(3):239–249

Hui-Li KT-FG (2005) Study of determination of organophosphorus pesticides using biosensors based on cobalt phthalocyanine polymer modified electrodes. Chem Sens 1:011

Ilangovan R, Daniel D, Krastanov A, Zachariah C, Elizabeth R (2006) Enzyme-based biosensor for heavy metal ions determination. Biotechnol Biotechnol Equip 20:184–189

Iqbal SS, Mayo SW, Bruno JG, Bronk BV, Batt CA, Chambers JP (2000) A review of molecular recognition technologies for detection of biological threat agents. Biosens Bioelectron 15 (11–12):549–578

Karim F, Fakhruddin ANM (2012) Recent advances in the development of a biosensor for phenol: a review. Rev in Environ Sci Biotechnol 11:261–274

Kim BH, Chang IS, Gil GS, Park HS, Kim HJ (2013) Novel BOD (biological oxygen demand) sensor using a mediator-less microbial fuel cell. Biotechnol Lett 25:541–545

Knecht MR, Sethi M (2009) Bio-inspired colorimetric detection of Hg^{2+} and Pb^{2+} heavy metal ions using au nanoparticles. Anal Bioanal Chem 394:33–46

Lang Q, Han L, Hou C, Wang F, Liu A (2016) A sensitive acetylcholinesterase biosensor based on gold nanorods modified electrode for detection of organophosphate pesticide. Talanta 156:34–41

Lee SA, Yang C (2014) A smartphone-based chip-scale microscope using ambient illumination. Lab Chip 14(16):3056–3063

Liang PS, Park TS, Yoon JY (2014) Rapid and reagentless detection of microbial contamination within meat utilizing a smartphone-based biosensor. Sci Rep 4:5953

Luzi E, Minunni M, Tombelli S, Mascini M (2003) New trends in affinity sensing. TrAC Trends Anal Chem 22:810

Mauriz E, Calle A, Manclús J (2007) Optical immunosensor for fast and sensitive detection of DDT and related compounds in river water samples. Biosens Bioelectron 22(7):1410–1418

Min WK, Cho YJ, Park JB, Bae YH, Kim EJ, Park K, Seo JH (2009) Production and characterization of monoclonal antibody and its recombinant single chain variable fragment specific for a food-borne mycotoxin, fumonisin B1. Bioprocess Biosyst Eng 33(1):109–115

Modin O, Wilén BM (2012) A novel bioelectrochemical BOD sensor operating with a voltage input. Water Res 46:6113–6120

Mostafa GA (2010) Electrochemical biosensors for the detection of pesticides. The Open Electrochem J 2:22–42

Mostafalou S, Abdollahi M (2013) Pesticides and human chronic diseases: evidence, mechanisms, and perspectives. Toxicol Appl Pharmacol 268(2):157–177

Mulchandani A, Mulchandani P, Keneva I, Chen W (1998) Biosensors for direct determination of organophosphate nerve agents using recombinant E. coli with surface expressed organophosphorus hydrolase. Anal Chem 70:4140–4145

Nicolini AM, Fronczek CF, Yoon JY (2015) Droplet-based immunoassay on a "sticky" nanofibrous surface for multiplexed and dual detection of bacteria using smartphones. Biosens Bioelectron 67:560–569

Nougadère A, Reninger JC, Volatier JL, Leblanc JC (2011) Chronic dietary risk characterization for pesticide residues: a ranking and scoring method integrating agricultural uses and food contamination data. Food Chem Toxicol 49(7):1484–1510

Okon SL, Ronkainen NJ (2017) Enzyme-based electrochemical glutamate biosensors. In: Electrochemist Sens tech. Intech Open Limited, London, pp 13–32

Pagkali V, Petrou PS, Makarona E, Peters J, Haasnoot W, Jobst G, Kakabakos SE (2018) Simultaneous determination of aflatoxin B 1,fumonisin B 1 and deoxynivalenol in beer samples with a label-free monolithically integrated optoelectronic biosensor. J Hazard Mater 359:445–453

Pal P, Bhattacharya D, Mukhopadhyay A, Sarkar P (2009) The detection of mercury, cadmium, and arsenic by the deactivation of urease on rhodinizedcarbon. Environ Eng Sci 26:25–32

Pisoschi A (2016) Potentiometric biosensors: concept and analytical applications-an editorial. Biochem Anal Biochem 5(3):e164

Peavy HS, Rowe DR, Tchobanoglous G (1988) Environmental Engineering. McGraw Hill Inc, Singapore

Pohanka M, Skládal P (2008) Electrochemical biosensors-principles and applications. J Appl Biomed 6:57–64

Podola B, Nowack ECM, Melkonian M (2004) The use of multiple-strain algal sensor chips for the detection and identification of volatile organic compound. Biosens Bioelectron 19:1253–1260

Prabhakar N, Arora K, Singh SP, Pandey MK, Singh H, Malhotra BD (2007) Polypyrrole-polyvinyl sulphonate film-based disposable nucleic acid biosensor. Anal Chim Acta 589(1):6–13

Reyes De Corcuera JI, Cavalieri RP (2003) Biosensors. In: Heldman D (ed) Encyclopedia of agricultural, food and biological engineering. CRC Press, Boca Raton, pp 119–123

Rogers KR, Gerlach CL (1996) Environmental biosensors – a status report. Environ Sci Technol 30:486–491

Rodriguez-Mozaz S, Reder S, Lopez de Alda M, Gauglitz G, Barceló D (2004) Simultaneous multi-analyte determination of estrone, isoproturon and atrazine in natural waters by the River ANAlyser (RIANA), an optical immunosensor. Biosens Bioelectron 19(7):633–640

Rodriguez BB, Borbot JA, Tothill IE (2004) Development of urease and glutamic dehydrogenase amperometric assay for heavy metals screening in polluted samples. Biosens Bioelectron 19 (1):1157–1167

Ruban G, Ruperd Y, Laveau B, Lucas E (2001) Self-monitoring of water quality in sewer systems using absorbance of ultraviolet and visible light. Wat Sci Tech 44:269–276

Sang S, Wang Y, Feng Q, Wei Y, Ji J, Zhang W (2015) The progress of new label-free techniques for biosensors: a review. Criticism 36:1–17

Sassolas A, Prieto-Simón B, Marty JL (2012) Biosensors for pesticide detection: new trends. Am J Analyt Chem 3(3):210

Schramm W, Paek SH, Voss G (1993) Strategies for the immobilization of antibodies. Immuno Methods 3(2):93–103

Schulze H, Vorlová S, Villatte F, Bachmann TT, Schmid RD (2003) Design of acetylcholinesterases for biosensor applications. Biosens Bioelectron 18(2–3):201–209

Shekhovtsova TN, Muginova SV, Bagirova NA (1997) Determination of organomercury compounds using immobilized peroxidase. Anal Chim Acta 344:145–151

Shukla P, Nigam V, Gupta R, Singh A, Kuhad RC (2013) Sustainable enzyme technology for environment: biosensors for monitoring of pollutants and toxic compounds. In: Kuhad RC, Singh A (eds) Biotechnology for environmental management and resource. Springer, New Delhi, pp 69–74

Silva LMC, Salgado AM, Coelho MAZ (2010) Agaricus bisporus as a source of tyrosinase for phenol detection for future biosensor development. Environ Technol 31:611–616

Soares JK, Morin KM, Mello C (2004) Antimicrobial peptides for use in biosensing applications. DTIC Document, Fort Belvoir

Soldatkin O, Kucherenko I, Marchenko S, Kasap BO, Akata B, Soldatkin A, Dzyadevych S (2014) Application of enzyme/zeolite sensor for urea analysis in serum. Mater Sci 42:155–160

Tauriainen S, Karp M, Chang W, Virta M (1998) Luminescent bacterial sensor for cadmium and lead. Biosens Bioelectron 13:931–938

Thévenot DR, Toth K, Durst RA, Wilson GS (1999) Electrochemical biosensors. Pure Appl Chem 71:2333–2234

Tombelli S, Minunni M, Mascini M (2005) Methods 37:48–56

Turner AP (2013) Biosensors: sense and sensibility. Chem Soc Rev 42:3184–3196

Van Dorst B, Mehta J, Bekaert K, Rouah-Martin E, De Coen W (2010) Recent advances in recognition elements of food and environmental biosensors: a review. Biosens Bioelectron 26:1178–1194

Velychko TP, Soldatkin OO, Melnyk VG, Marchenko SV, Kirdeciler SK, Akata B, Soldatkin AP, El'skaya AV, Dzyadevych SV (2016) A novel Conductometric urea biosensor with improved analytical characteristic based on recombinant urease adsorbed on nanoparticle of Silicalite. Nanoscale Res Lett 11:106

Verma N, Bhardwaj A (2015) Biosensor technology for pesticides – a review. Appl Biochem Biotechnol 175:3093–3119

Nigam VK, Shukla P (2015) Enzyme based biosensors for detection of environmental pollutants-*A Review*. J Microbiol Biotechnol 25(11):1773–1781

Wang K, Gan L, Jiang L, Zhang X, Yang X, Chen M, Lan X (2015) Neutralization of staphylococcal enterotoxin B by an Aptamer antagonist. Antimicrob Agents Chemother 59(4):2072–2077

Wang L, Liu X, Zhang Q (2012b) Selection of DNA aptamers that bind to four organophosphorus pesticides. Biotechnol Lett 34(5):869–874

Wang X, Lu X, Chen J (2014) Development of biosensor technologies for analysis of environmental contaminants. Trends Environ Anal Chem 2:25–32

Wang Y, Zhang X, Zhang C, Liu Y, Liu X (2012a) Isolation of single chain variable fragment (scFv) specific for Cry1C toxin from human single fold scFv libraries. Toxicon 60 (7):1290–1297
Wei Q, Qi H, Luo W, Tseng D, Ki SJ, Wan Z, Ozcan A (2013) Fluorescent imaging of single nanoparticles and viruses on a smart phone. ACS Nano 7(10):9147–9155
Xiao Y, Pavlov V, Gill R, Bourenko T, Willner I (2004) Lighting up Biochemiluminescence by the surface self-assembly of DNA-Hemin complexes. Chem Bio Chem 5(3):374–379
Xu ZX, Gao HJ, Zhang LM, Chen XQ, Qiao XG (2011) The biomimetic immunoassay based on molecularly imprinted polymer: a comprehensive review of recent progress and future prospects. J Food Sci 76(2):R69–R75
Yazgan I (2014) Biosensor for selective detection of E. coli in spinach using the strong affinity of derivatized mannose with fimbrial lectin. Biosens Bioelectron 61:266–273
Yu H, Tan Y, Cunningham BT (2014) Smartphone fluorescence spectroscopy. Analytic Chem 86 (17):8805–8813
Yu Y, Liu Y, Zhen SJ, Huang CZ (2013) A graphene oxide enhanced fluorescence anisotropy strategy for DNAzyme-based assay of metal ions. Chem Comm 49(19):1942
Zangheri M, Cevenini L, Anfossi L, Baggiani C, Simoni P, Di Nardo F, Roda A (2015) A simple and compact smartphone accessory for quantitative chemiluminescence-based lateral flow immunoassay for salivary cortisol detection. Biosens Bioelectron 64:63–68
Zhang C, Wang L, Tu Z (2014) Organophosphoruspesticides detection using broad-specific single-stranded DNA based fluorescence polarization aptamer assay. Biosens Bioelectron 55:216–219
Zhi-tao C, Yan Z, Xiang-Ping Z, Jin-Hu Z, Wen-Fan W, Jian-Hui Z (2010) Determination of the phosalone by flow-injection chemiluminescent. Fujian Anal Test 4:007

Nanobiosensor in Health Sector: The Milestones Achieved and Future Prospects

4

Rubia Noori, Razi Ahmad, and Meryam Sardar

Abstract

A biosensor is a measurement device for the detection of an analyte that contains a biological material with a signal transducer. Current advances in nanotechnology and nanomaterial synthesis have created a new biosensor called nanobiosensor. In this sensor, the biological molecule is immobilized on the nanomaterial to form a compact probe. The reaction between the biomolecule and the analyte is heterogeneous in nature; therefore, the surface of this biointerface is very crucial in the performance of the nanobiosensor. To further improve their performance, various kinds of nanomaterials have been designed. Due to their large surface area, nanomaterials possess high sensitivity and thus, enhance its application in the detection and diagnosis of various diseases. This chapter provides an overview of various types of nanobiosensors that have been developed for biological, environmental, and medical applications. Different types of biosensors fabricated using various biomolecules are discussed in detail and a brief idea on the mode of transducer reaction is also highlighted. The later part of the chapter gives information about applications of nanobiosensor in diverse fields, mostly in the health sector.

Keywords

Nanobiosensor · Nanomaterials · Transducer · Nanoparticles · Carbon nanotubes · Biosensing devices

R. Noori · R. Ahmad · M. Sardar (✉)
Department of Biosciences, Jamia Millia Islamia, New Delhi, India
e-mail: msardar@jmi.ac.in

M. Mohsin et al. (eds.), *Nanobiosensors for Agricultural, Medical and Environmental Applications*, https://doi.org/10.1007/978-981-15-8346-9_4

Abbreviations

MNPs	Magnetic nanoparticles
AgNPs	Silver nanoparticles
Au	Gold nanoparticles
QDs	Quantum dots
CNTs	Carbon nanotubes
NW	Nanowires
NR	Nanorods
SiNW	Silicon nanowire
FET	Field-effect transistor
SPR	Surface plasmon resonance
LSPR	Localized surface plasmon resonance
SELEX	Systematic evolution of ligands by exponential enrichment
DMR	Diagnostic magnetic resonance
GNRs	Gold nanorods
MEMS	Microelectromechanical system
NEMS	Nanoelectromechanical systems
ml	Microliter
mM	Millimolar

4.1 Introduction

The biosensor is an analytical device that senses biological materials and estimates it by biological signals. These signals are then evaluated and converted into readable form using transduction and electromechanical interpretation. Figure 4.1 represents a model of a biosensor and gives information about its components. There are three major components of a biosensor, namely bioreceptor, transducer, and detector. Their main function is to sense a biological material like immunological molecules, biomolecules, and enzymes.

Nanomaterials have unique properties; therefore, they can be exploited for the development of sensitive nanobiosensor. The nanomaterial-based sensors are economical, reasonable, robust, and reproducible (Rai et al. 2012). Thus, nanobiosensor is used to determine a biochemical or biological event through a compact probe

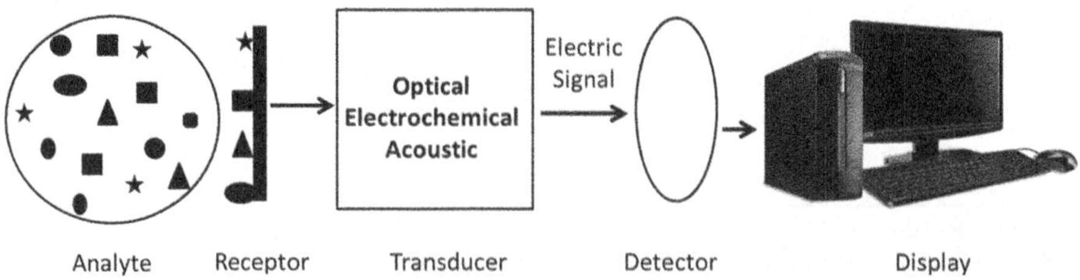

Fig. 4.1 Graphical representation of component of biosensor

(Di Giusto et al. 2005; Gullberg et al. 2004; Prasad 2014). At present, it is considered that nanobiosensor technology can revolutionize the health care industry because of applications for human health like monitoring of diabetes, measurement of metabolites, detection of cancer biomarkers, and in forensic medicine, etc. In the victuals industry, nanobiosensor is used for remote sensing of water quality, monitoring the presence of drug residues in food. They can also be used for the detection of pesticides in the environment, heavy metals in river water, and genome analysis of organisms. The nanomaterials like silver, gold, magnetic nanoparticles, carbon nanotubes, and quantum dots have been vigorously investigated for their application as biosensors. Several studies reported that the silver and other metal nanoparticles have significant applications in the field of biolabeling (Kulesza et al. 2009), drug delivery system (Lv et al. 2009; Marcato and Durán 2008), filters, and as antimicrobial drugs (Durán et al. 2010; Rai et al. 2009) and sensors (Baruah and Dutta 2009).

4.2 Nanobiosensor and Their Properties: Fusion of Nanotechnology with Biosensor

Nanotechnology has contributed a lot in the field of biosensor development due to its unique electronic, magnetic, and optical properties of the nanomaterials used. The special characteristics like submicron size and large surface area to volume ratio enhance the sensitivity of the nanosensor for the analyte. Nanomaterials are considered environmentally sustainable and thus can be utilized in sensing technology. Nanobiosensor is defined as a device that is used to detect analyte quantitatively with the help of a biologically active element attached to a suitable nanostructured transducer (Fan et al. 2008; Velasco-Garcia 2009; You et al. 2009). Nanobiosensor is a modified version which has biologically sensitive molecule immobilized onto the surface of the physico-chemical transducer. It is the effort of researchers working in diverse fields like electronics, biologists, material chemist, and physicists (Turner 2000). It is a compact analytical device that has enhanced inherent specificity, quick response time, and reliability. The important property of a biosensor is the specificity and it should be high enough to determine analyte from the rest of the unwanted material. The interaction between the target analyte and sensor should not be affected by physical factors like temperature and pH. Also, the sensor should provide analysis results with precision, accuracy, and linearity without the disturbance of electrical noise. Stability is another important criterion for an ideal nanobiosensor under normal storage conditions. It should be economical, inexpensive, manageable and should be conveniently used by semi-skilled operators (Rai et al. 2012). Although nanobiosensor is still in the developing stage but it has shown promising results in bioanalytical applications and has gained popularity because of its ability to detect ultra-low concentrations of any analyte that are observed to be a potent threat to the living being. It works at the atomic level with great efficiency, and due to its high sensitivity, sometimes it is considered to be an error-prone system (Rai et al. 2012).

The biologically active or recognition element, the transducer, and the detector are three important components to formulate any nanobiosensor. The biological elements used are bioreceptor probes that are highly selective for target analytes like antibodies, nucleic acids, pathogens, and metabolites. Molecular recognition elements include biologically derived materials or bio-mimic components such as receptors, enzymes, nucleic acids, antibodies, molecular imprints, lectins, tissue, microorganisms, and organelles (Razavi and Janfaza 2015; Sharma and Rogers 1994). The working of nanobiosensor starts with the binding of targeted bioanalytes with the bioreceptors and this binding generates a physicochemical signal which is modulated by the receptor. The transducer can be electrochemical, mass, optical, and thermal. The transducer measures changes which occur as a result of heterogeneous reaction at bioreceptor/element and thus acts as a borderline between receptor and detector. It captures and transforms the signal into measurable electrical output. This electrical signal is trapped by the detector component which is further amplified and observed by the microprocessor (Prasad 2014).

In nanobiosensor, the change in the signal like electric potential, current, impedance, intensity, current, and phase of electromagnetic radiation is measured. These variations are analyzed and confirm the presence or absence of bioanalytes. The properties of transducers and biorecognition element determine the sensitivity of a nanobiosensor. The nanostructured materials are located between the biological agents and detector component and thus behave as intermediary phase and this nanomaterial is immobilized on the transducer in order to fabricate a nanobiosensor. Several nanomaterials have been screened for the development of nanobiosensor (Gomes et al. 2015; Sharma and Rogers 1994). The nanostructured materials of different chemical nature are discussed in the next section.

4.2.1 Nanomaterials Used in Biosensing Devices

Recent years have seen tremendous growth in using nanoscale materials for developing electrochemical biosensing devices as the properties of nanomaterials offer an excellent outlook in constructing high performance, novel sensing systems. The nanomaterials are incorporated into the transducer which can send the observable signal to the recorder. Controlling the size and morphology of nanomaterials can enhance the power of detection, sensitivity, and transducing capability to a certain degree (Pak et al. 2001). There are few factors which are kept in mind while selecting a particular kind of nanoscale material.

Materials at the nanoscale level are integrated with highly sensitive electrical and electromechanical properties when engineered with the nanoelectromechanical systems (NEMS). It has enabled the materials to gain complex electrical, mechanical, fluidic, thermal, optical, and magnetic properties. NEMS is a class of devices and as the name suggests, the device is in the nanometer range and thus having novel features like low mass, high mechanical resonance frequencies, more surface-to-volume ratio, and large quantum mechanical effects such as zero-point motion. NEMS and MEMS (microelectromechanical system) devices have resulted in mechanical materials to perform in a much better and sophisticated way as the mechanical property of a material determines its size. MEMS is a mini device with electrical and mechanical components. Using these devices, surface forces like adhesion, cohesive, friction, and viscous drag force can be regulated precisely which helps in biochemical interaction taking part in sensing technology. Incorporating nanomaterials with these devices increases the response to a wide range of stimuli (Bhushan 2007).

Optimizing the optical properties of nanomaterials is also one of the important criteria considered for its selection. Plasmon surface resonance is an interesting feature of nanomaterials which maximizes the optical response of the sensing materials with the incident light. Ionic and charged species are used to excite the surface of sensing materials and cause excitation of the fluidic state of charged particles. Due to this phenomenon, nanoparticles have photonic character and can be used as fluorophores. The refractive index of a medium is a vital property that governs the flow of light through a medium and also affects the surface plasmon resonance. A nanobiosensor is thus able to detect the infinitesimal biological interactions and results in a much better and reliable degree of estimation. So, for the implementation of nanomaterials, they are to be optimized for their performance and effect to be used in biosensing devices as per the required goal (Haes and Van Duyne 2002; Kelly et al. 2003). Tailoring of physical/chemical properties like shape, size, structure, and composition of nanomaterials is done to alter the specific absorptive, emissive, and light-scattering features. Various nanostructured materials have been modified and utilized with specific forms such as 0D (quantum dots, nanoparticles), 1D (nanowires or carbon nanotubes), or 2D (metallic platelets or graphene sheets) orientation that reflects in their final properties (Pandit et al. 2016). These nanomaterials are described below.

4.2.1.1 Metallic Nanoparticles

Due to their microscopic size, high surface to volume ratio, electrical and optical properties, metallic nanoparticles have been a huge success in diverse fields. They have been exploited in bioanalytical applications for the development of biosensors, diagnostics, imaging, drug delivery, and therapy. Among metal NPs (1–100 nm in size), gold nanoparticles (GNPs) have been widely employed for such applications. They have a large surface area, strong adsorption ability, scattering properties, and facile conjugation to various biomolecules and are considered as nontoxic, biocompatible, and inert core nanomaterials. Gold NPs are being extensively used for the detection of various analytes by fabricating it into immunoassays, diagnostics, and biosensors. They can act as nanocarriers for the delivery of drugs, DNA, and genes in the therapy of cancer and other diseases (Kim et al. 2002; Park et al. 2013). After GNPs, magnetic NPs (MNPs) are the second most common NPs which have been used in the development of nanobiosensor and detection of analytes such as proteins, enzymes, DNA, mRNA, drugs, metabolites, pathogens, and tumor cells. The MNPs are being extensively used by industries to develop diagnostic magnetic resonance (DMR) technology and nuclear magnetic resonance detector (μNMR) which can detect a sensitive analyte in microliter volumes (Sahoo et al. 2017). Silver nanoparticles (AgNPs) are also very commonly synthesized and considered to be a noble metal having attractive physicochemical properties including the surface plasmon resonance and large effective scattering cross-section of individual silver nanoparticles. They have been used in diverse applications effectively including the detection of biological macromolecules. Hydrophobic Ag–Au composite nanoparticles have been observed to show strong adsorption and good electrical conducting properties and thus are being used in biosensing devices (Link et al. 1999; Rai et al. 2012).

4.2.1.2 Carbon Nanotubes (CNTs)

Since the discovery of carbon nanotubes (CNTs) in the 90s, its application has been heavily investigated in biosensing devices. The first CNT-based nanosensor was fabricated in 2003 by Wang and Musameh (2003) and Yun et al. (2009). CNTs are an exciting, one-dimensional, and a new form of carbon-based nanostructured material that has already been exploited in many fields like diagnostics, tissue engineering, cell tracking and labeling, and delivery of drugs and biomolecules (Pandit et al. 2016). CNTs are cylindrical in shape and consist of one, two, or several concentric graphene sheets seamlessly wrapped into a tube, capped by fullerenic hemisphere. There are two main types of CNTs—single-walled and multi-walled. CNTs have gained attention due to their unique properties such as structure-dependent electronic and mechanical properties, high thermal conductivity, excellent biocompatibility, high chemical stability, extraordinary electrocatalytic activity, very low surface fouling, low overvoltage, and high surface to volume ratio (Pandit et al. 2016). The surface of CNTs can be modified by adsorbing biomolecules such as protein, DNA, etc., electrostatically or it can be attached to the functional group present on the CNTs. CNT-based nano biosensor has been used in the diagnosis of analytes in healthcare, industries, environmental monitoring, and food quality analysis.

4.2.1.3 Graphene

Graphene is made up of a thin layer of sp2-hybridized carbon. Due to its exquisite properties such as electron transfer ability, high mechanical strength, high thermal conductivity, tunable optical properties, tunable bandgap, high elasticity, very high room temperature electron mobility, and demonstration of the room temperature quantum Hall effect, graphene is becoming more popular than other nanomaterials in bioanalytical and bioimaging applications. Their superior performance facilitates them to be widely used in electrochemical, impedance, fluorescence, and electrochemiluminescence biosensors for the detection of a wide range of analytes such as glucose, cytochrome c, NADH, hemoglobin, cholesterol, ascorbic acid, dopamine, uric acid, hydrogen peroxide, horseradish peroxidase, catechol, DNA, heavy metal ions, and gases. They also have low production costs and minimize harmful environmental effects (Pandikumar et al. 2014).

4.2.1.4 Quantum Dots (QDs)

QDs have unique spectral properties and thus recently, they have been exploited as a new generation of fluorophores in bioimaging and biosensing. They are nanostructured materials with size ranging from 1 to 10 nm. QDs have unique optical properties of broad excitation, narrow size-tunable emission spectra, high quantum yield and molar extinction coefficients, high photochemical stability, and negligible photobleaching (Androvitsaneas et al. 2016). Optical biosensors have been developed using QDs (inorganic nanocrystals) as nanomaterials exclusively for the detection of ions, organic compounds, pharmaceutical analytes, and biomolecules such as nucleic acids, proteins, amino acids, enzymes, carbohydrates, and neurotransmitters (Androvitsaneas et al. 2016; Bakalova et al. 2004; Bulovic et al. 2004).

4.2.1.5 Nanowires and Nanorods

Nanowires are semi-conductor nano-structured materials having unique optical and electronic properties with size ranging from several tens to over 100 μm length and are sensitive enough to detect the binding events of small molecules, peptide nucleic acid (PNA)–DNA as well as DNA–DNA hybridization. Due to their structure, nanowires also show distinctive absorption and photocurrent characteristics. Nanowire-based nanobiosensor shows the sensing procedure when there is a change in charge density (inducing the change in the electric field on the nanowire surface) upon binding of biomolecules (negatively charged molecules bind to the n-type field-effect transistor) (Ambhorkar et al. 2018; Panpradist and Lai 2016).

Nanorods are nanoscale objects having dimensions range of 1–100 nm. Due to their optoelectronic properties, they are gaining momentum in developing different designs of nanosensors. Gold nanorods (GNRs) are considered an excellent candidate to be used in sensing devices as the absorbance band changes with the refractive index of local material and thus allows extremely accurate sensing. GNRs were used in detecting target sequences of infecting agents of many dangerous diseases, for example, an HIV-1 (He et al. 2008).

4.3 Types of Nanobiosensor

Nanobiosensor has been classified broadly based on bioreceptor used and the types of the transducer. Different biomolecules are being immobilized on the surface of the transducer as a bioreceptor which can be divided into several classes such as enzymatic biosensor, DNA/RNA biosensor, immunosensor, aptasensor, and microbial biosensor (Razavi and Janfaza 2015). The classification can also be done based on transducer used for sensing such as electrochemical, optical, etc.

4.3.1 Classification Based on Bioreceptor

4.3.1.1 Enzymatic Biosensor

Enzymes that are specific for an analyte to convert it into a product can be immobilized with a suitable transducer. Figure 4.2 represents a schematic diagram of an enzymatic biosensor. Enzymatic biosensor measures the activity of enzymes selectively upon interaction with a specific target and generates a biological signal proportional to the target analyte concentration. These biosensors have been used to detect analytes at nanoscale level especially with an electrochemical transducer. There is a heterogeneous electron transfer occurring between the electrode and the protein redox center which governs the performance of this biosensor. The enzymatic biosensor is the most common nanobiosensor developed till now, for e.g., glucose nanobiosensor for the rapid self-diagnosis of blood glucose levels (Lad et al. 2008; Trojanowicz 2002; Wang et al. 2014).

4.3.1.2 Oligonucleotide (DNA/RNA) Biosensor

Oligonucleotide biosensors are diagnostic devices that consist of a probe (single-stranded DNA/RNA) that is in association with or integrated within a transducer or transducing micro nanosystem as shown in Fig. 4.3. The transducer employed for achieving high sensitivity can be electrochemical, optical, thermometric, piezoelectric, magnetic, or micromechanical (del Valle and Bonanni 2014). DNA based biosensor is a current approach and researchers are trying to develop DNA biosensors for low-cost detection of specific DNA sequences in human, viral, and bacterial nucleic acids, several attempts have been made to increase sensitivity and selectivity of the sensor (Zhao et al. 2014). Nanomaterials play an important role in DNA biosensor, i.e. it is used as a substrate for DNA attachment as well as amplifies the signal for hybridization, an enhanced amount of DNA immobilization occurs on nanomaterial and another advantage is that DNA maintains its biological activity. Nanoparticles like gold, cadmium sulfide, nanowires like silicon, nanotubes like carbon nanotubes, etc., are being employed for DNA biosensor (Sheehan and Whitman 2005; Shi et al. 2013). Electrochemical based DNA biosensor has been

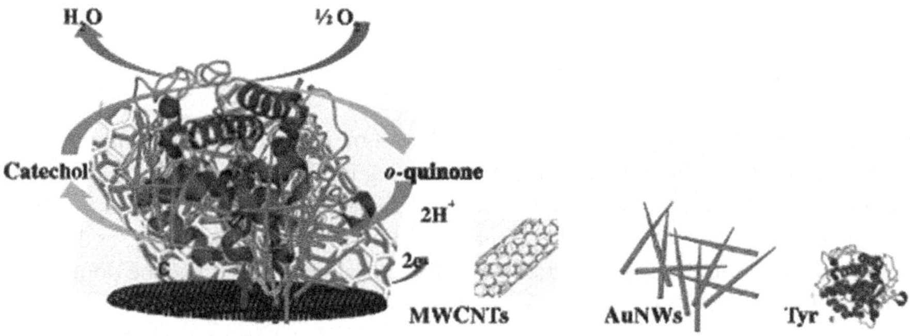

Fig. 4.2 Schematic representation of the enzymatic biosensor. Reprinted with permission from Kurbanoglu et al. (2017). © 2018, Springer

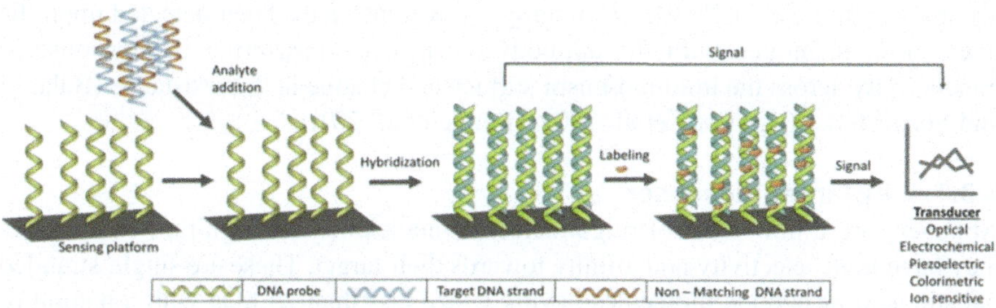

Fig. 4.3 General design of DNA biosensor. Reproduced from Asal et al. (2018). © 2018 (CC BY 4.0)

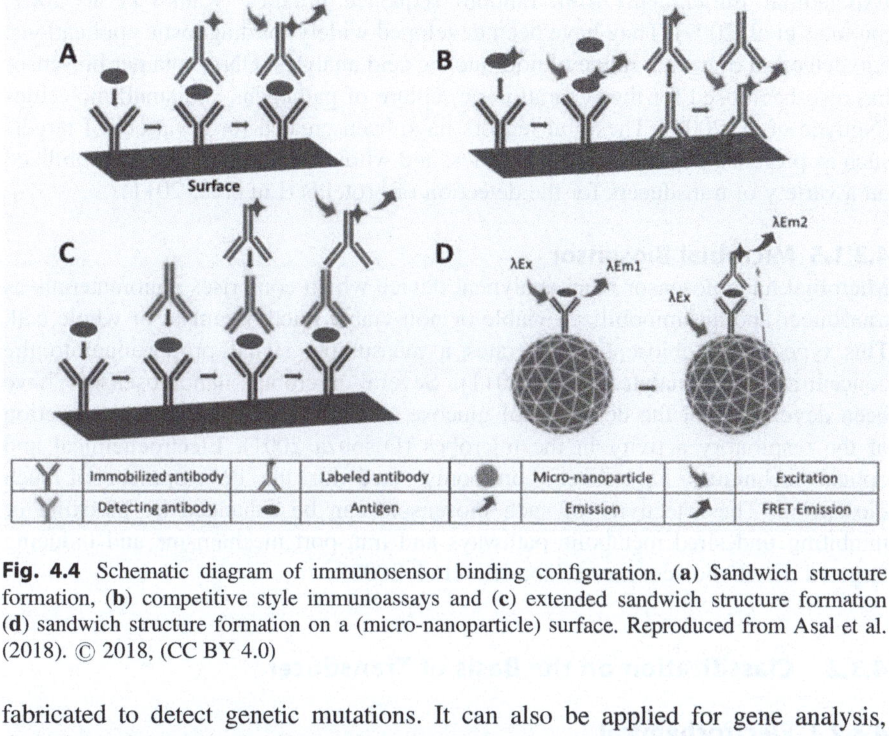

Fig. 4.4 Schematic diagram of immunosensor binding configuration. (**a**) Sandwich structure formation, (**b**) competitive style immunoassays and (**c**) extended sandwich structure formation (**d**) sandwich structure formation on a (micro-nanoparticle) surface. Reproduced from Asal et al. (2018). © 2018, (CC BY 4.0)

fabricated to detect genetic mutations. It can also be applied for gene analysis, detection of genetic disorders, clinical diagnostics, tissue matching, and forensic investigation (Odenthal and Gooding 2007).

4.3.1.3 Immunosensor

Immunosensors are based on immobilized antibodies as a biological recognition element that is highly specific and has been widely applied in clinical analysis; a typical immunosensor is shown in Fig. 4.4. These have been generally used to detect disease at a molecular level. Microorganisms like *Escherichia coli*, *Salmonella*, *Staphylococcus aureus*, pesticides, herbicides, etc., have been detected with high accuracy (Shirale et al. 2010). Viral antigens of tumor have been detected upon the interaction of antigen with the antibody receptor by measuring the response in conductivity across the immunosensor surface and change in the resistance (Bahadır and Sezgintürk 2015; Cruz et al. 2002; Shirale et al. 2010).

4.3.1.4 Aptamer Biosensor

Aptamers are a new class of oligonucleotide/nucleic acid recognition elements as they have high selectivity and affinity towards their target. These are single-stranded nucleic acid or peptide molecules having a size less than 25 kDa with a natural or synthetic origin. They can be used for the development of a sensor for DNAs, proteins, and small molecules. The nucleic acid molecules are selected by an in vitro selection process called SELEX (Systematic Evolution of Ligands by Exponential Enrichment) from random sequence libraries (Radko et al. 2007; Sassolas et al. 2009). They have been developed widely for diagnostic applications, e.g. detection of a wide range of non-nucleic acid analytes. DNA aptamer biosensor has also been used for the separation or capture of pathogens and small molecules (Nguyen et al. 2009). These biosensors have been created for a variety of targets such as proteins, peptides, small organics, and whole cells. It has been immobilized on a variety of transducers for the detection of proteins (Liu et al. 2011).

4.3.1.5 Microbial Biosensor

Microbial nanobiosensor is an analytical device which comprises nanomaterials as transducer and an immobilized viable or non-viable microorganism or whole cell. This type of nanobiosensor generates a measurable signal proportional to the concentration of analytes (Shin 2011). Several microbial nanobiosensors have been developed for the detection of glucose on the basis of oxygen consumption of the respiratory activity in the microbes (D'souza 2001). Electrochemical and optical techniques have been commonly used for the development of such biosensors. The selectivity of such biosensors can be enhanced by blocking or inhibiting undesired metabolic pathways and transport mechanisms and inducing targeted metabolic activities (Gäberlein et al. 2000).

4.3.2 Classification on the Basis of Transducer

4.3.2.1 Electrochemical

Most of the nanobiosensors developed so far are based on electrochemical detection as they have many advantages over other sensing techniques such as fast response time, highly sensitive, low cost, low-interference characteristics, and compatible with microfabrication technology, small size, economical cost, minimum power requirement, ease of use, and low maintenance (Bertók et al. 2013; Lad et al. 2008). In this technique, biochemical reactions between the nanofabricated biomolecule or biological element and target analyte are being analyzed with the help of electrical means (Chaubey and Malhotra 2002; Cai et al. 2001). It works on the principle of electrochemistry. The readable electrochemical signals are detected during a bio interaction when an electron is consumed or produced and are measured by the electrochemical method (Chaubey and Malhotra 2002). The electrical signals quantitatively correspond with the concentration of analyte present in the sample (Ronkainen et al. 2010). In a study by Sistani et al. (2014), biochemical detection of penicillin was done successfully using penicillinase enzyme immobilized on nanoparticles (Sistani et al. 2014). Electrochemical nanosensors can be categorized into potentiometric, conductometric, amperometric, and impedimetric biosensors according to their working principle.

Potentiometric nanosensor is based on measuring the potential of a system at a working electrode with respect to a sensitive and accurate reference electrode under zero current flow (Koncki et al. 2000). This potential is generated by converting the biorecognition process and obtaining analytical information of a system (Dzyadevych et al. 2004). The potential signal is generated due to the accumulation of ions at ion-selective electrodes and ion-sensitive field-effect transistors at equilibrium (Koncki et al. 2000). It has been well-established and utilized in the biomedical field, in the detection of various analytes such as antibiotics, preservatives, heavy metals, and pesticides in agricultural and food industries (Durán and Marcato 2013).

Conductometric nanosensor measures change in conductivity of the solution once the interaction of target and immobilized analyte is achieved. The principle is based on a change occurring in electrical resistance between two parallel electrodes during a biochemical reaction (Mikkelsen and Rechnitz 1989; Muhammad-Tahir and Alocilja 2003).

Amperometric nanosensor allows subsequent measurement of current produced by oxidation or reduction of an electroactive species in an electrochemical reaction. They have been mostly used in medical devices and offer many advantages over other electrochemical sensors as they are highly sensitive, fast, precise, accurate, and economical and have a wide linear range (Wang 1999). It has been used in fields of health and diagnostics like development of ATP sensor (Kueng et al. 2004) and beta-HCG sensor (Santandreu et al. 1999) for pregnancy test, in environmental and agriculture like detection of organophosphates, ractopamine, sulfonamides, and hydrogen peroxide (Lin et al. 2013; Xu et al. 2013; Yan et al. 2013).

Impedimetric nanosensors have not been used frequently as compared to other electrochemical sensors. It basically measures the electrical impedance of a particular biological system and gives analytical information about that system (Chuang et al. 2011; Huang et al. 2008).

4.3.2.2 Optical

Optical nanobiosensor measures the change in optical signal and this change is recorded in resonant frequency after the interaction between the analyte and a resonator which oscillates a light within a cavity (Vo-Dinh 2005). They are a powerful and versatile detection tool and highly sensitive to biomolecular targets and provide a quick response. They are insensitive to electromagnetic interference. Optical methods include surface plasmon resonance (SPR), localized surface plasmon resonance (LSPR), surface-enhanced Raman scattering, fluorescence spectroscopy, colorimetric spectroscopy, total internal reflectance, light rotation, and polarization (Borisov and Wolfbeis 2008).

Surface plasmon resonance (SPR) is an optical sensing technique that involves the interaction of light with the electrons of metal and hence causes transfer of energy to electrons present at the surface of the metal (Haes and Duyne 2004). SPR based biosensors have been successfully utilized in fundamental biological studies, drug discovery, health science research, and clinical diagnosis and can detect a wide range of analytes such as proteins, small molecules, antibody–antigen, DNA and RNA hybridization, concanavalin A, antibiotics, mycotoxins, and pathogen like *E. coli* (Liedberg et al. 1983).

In localized surface plasmon resonance (LSPR), a local oscillation occurs in the close proximity of metallic nanoparticles (MNPs). Light interacting with the MNPs are much smaller than the incident wavelength. This technique has been utilized successfully due to its label-free method as they do not require labeling of the target molecule with any kind of reagent and high sensitivity favors the method (Jia et al. 2012).

Fluorescence biosensor measures change in fluorescence which recognizes and gives information on the presence, activity, or conformation of a given target specifically and quantitatively and hence provides the dynamic molecular behavior. They are known to have high sensitivity, quick response and have the ability to achieve high spatial resolution through spectroscopic and imaging methods. Detection of some analytes such as nitrite, reactive oxygen species, pathogenic bacteria such as *S. aureus, V. parahaemolyticus*, *S. typhimurium,* and *E. coli* have been done by this technique (Chen et al. 2016; Dasary et al. 2008; Hu et al. 2014; Wu et al. 2014).

Colorimetric biosensor measures the change in absorption when a reaction proceeds and forms a colored product. It has been applied to a diverse field such as environmental detection of the toxic metal, clinical diagnosis of analytes, such as glucose, cancer biomarkers, and viruses (Cao et al. 2014; Chen et al. 2014; Sener et al. 2014; Zhang et al. 2014).

4.3.2.3 Piezoelectric

The piezoelectric sensor measures the change in the resonant frequency of piezoelectric quartz which oscillates under the effect of electric field and this frequency is proportional to surface adsorbed mass which changes due to absorption or desorption of molecules from the surface of the quartz. The crystals contain biorecognition element on its surface which selectively binds with the target and causes a change in mass. It has been applied for the detection of viruses, bacteria, proteins, and nucleic acids due to their great sensitivity and selectivity (Borman 1987; Durmuş et al. 2015; Tilmaciu and Morris 2015).

4.4 Applications of Nanobiosensor

Nanobiosensor has diverse applications in various fields like biomedical, environmental, and agricultural industry, etc., as highlighted in Fig. 4.5. Blood glucose sensing is the most common and major application in biosensing due to its plentiful market potential. Biosensors have incredible market value in other fields, but commercial adoption was slow due to several technological difficulties. Biosensor contamination is a major problem, due to the presence of biomolecules along with semiconductor materials.

Current developments in the field of nanoelectronics, biological, and information technology present way to develop medical nanorobots which are implanted and incorporated devices. The nanorobotics consists of main sensing, actuation, data transmission, remote control uploading, and coupling power supply subsystems (Appenzeller et al. 2002; Cavalcanti et al. 2007; Freitas 1999; Lavan et al. 2003; Liu and Shimohara 2007; Shi et al. 2007). Nanorobots should offer new tools for common medical treatments because it has nanoscopic quality (Leary et al. 2006; Patel et al. 2006). Table 4.1 lists different nanomaterials used for fabricating nanobiosensor with various sensing techniques and applications in diverse fields.

4.4.1 Detection of Glucose/Other Metals

Conventional methods are often tiresome in diabetes detection and monitoring. To eliminate this problem, the nanorobot sensor should be used to sense biological changes linked with hyperglycemia as it generates proteomic-based information (Cavalcanti et al. 2008). This will also make it possible to treat diabetes more quickly and effectively (Cash and Clark 2010; Gordijo et al. 2011; Samuel et al. 2010). Recently, Yang et al developed a very sensitive colorimetric detection method to detect early diabetes analysis by glucose detection in urine as an

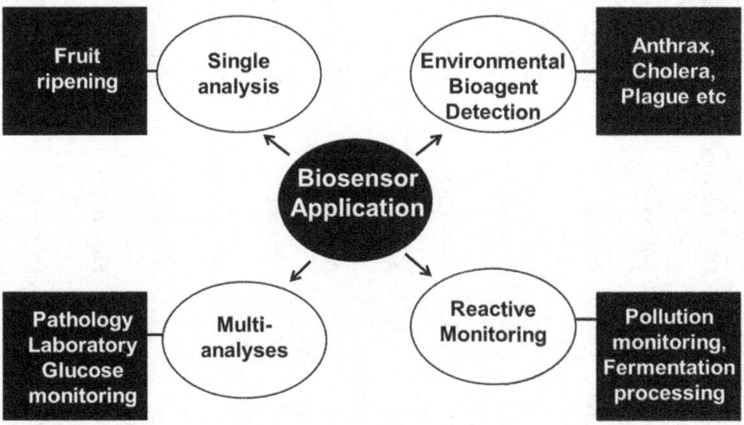

Fig. 4.5 Diverse applications of nanobiosensor

Table 4.1 Table summarizes different nanomaterials used for fabricating nanobiosensor

Nanomaterials	Recognition element	Target element	Sensing technique	Application	References
Gold nanoparticles	Oligonucleotide	*Listeria monocytogenes* and *Salmonella enterica*	Colorimetric	Food quality analysis	Devi et al. (2013)
Gold nanoparticles	Antibodies	*Salmonella typhimurium*	Surface plasmon resonance immunosensor	Clinical diagnostics and environmental monitoring	Ko et al. (2009)
Magnetic nanoparticles	Aptamer	*Staphylococcus aureus*, *Vibrio parahaemolyticus*, and *Salmonella typhimurium*	Luminescence	Food safety analysis	Wu et al. (2014)
Silica nanoparticles	Antibody	*Escherichia coli*	Fluorometric	Detect pathogen in ground beef sample	Zhao et al. (2004)
Gold nanoparticles	Antibiotics	Neomycin, kanamycin, and streptomycin	Surface plasmon resonance	Analyze antibiotics in milk sample	Frasconi et al. (2010)
Gold nanoparticles	Antibody	Aflatoxin M1	Dynamic light scattering	Mycotoxins detection in milk samples	Zhang et al. (2013)
Gold nanoparticles	Aptamer	Malathion	Surface-enhanced Raman spectroscopy	Detection of pesticide	Barahona et al. (2013)
Gold nanoparticles	Aptamer	Acetamiprid	Optical	Detection of pesticides	Weerathunge et al. (2014)
Gold nanoparticles	Oligonucleotide	Cytochrome b (cytb) gene	Fluorometric	Pork adulteration in processed mixed meat products	Ali et al. (2011)
Silver nanoparticles	Antibody	Myc-tagged protein	Optical	Detect protein molecule biology	Cao et al. (2009)
Silver nanoparticles	Enzyme	Penicillin	Electrochemical	Detection of molecules for clinical, biomedical analysis	Sistani et al. (2014)
Gold nanoparticles	Aptamer	Malathion	Surface-enhanced Raman spectroscopy	Detection of pesticide	Barahona et al. (2013)

Gold nanoparticles	Oligonucleotide	Tumor suppressor gene, adenomatous polyposis coli (APC)	Fluorometric	To determine the concentration of APC gene in human plasma sample	Darestani-Farahani et al. (2018)
Graphene oxide and iridium oxide nanoparticle	Enzyme	Angiotensin-converting enzyme inhibitor drug, captopril	Electrochemical	Determination of drug in spiked human serum and pharmaceutical dosage	Kurbanoglu et al. (2017)
Platinum nanoparticles	Antibody	Prostate specific antigen (PSA)	Electrochemical	Prostate cancer diagnosis	Spain et al. (2016)
Iron oxide nanoparticles	Oligonucleotide	microRNA (miRNA)	Surface-enhanced Raman scattering	miRNA related cancer diagnosis	Pang et al. (2016)
Silica and gold nanoparticles	Oligonucleotides	microRNA-21 and 141 (miRNA)	Electrochemiluminescence and voltammetric	miRNA biomarkers detection in a clinical laboratory	Feng et al. (2016)
Gold nanoparticles	Aptamer	Thrombin	Optical	Detection of thrombin in solution and on surfaces	Pavlov et al. (2004)
Carbon nanotubes	Enzyme	Glucose	Electrochemical	Detection of glucose in a variety of biological fluids (e.g., saliva, sweat, urine, and serum)	Lin et al. (2004)
Carbon nanotubes	Enzyme	L-Lactate	Electrochemical	Determination of lactate in commercial embryonic cell culture medium	Hernández-Ibáñez et al. (2016)
Carbon nanotubes	Antibody	Survival motor neuron (SMN) 1 gene	Electrochemical	Detection of Spinal muscular atrophy	Eissa et al. (2018)
Carbon nanotubes	Aptamer	RAP1 GTPase and HIV integrase	Optical	Detection of individual proteins from *Escherichia coli* (bacteria) and *Pichia pastoris* (yeast)	Landry et al. (2017)
Carbon nanotubes	Aptamer	Kanamycin	Fluorometric	Detection of antibiotic kanamycin in standard solutions as well as in milk samples	Liao et al. (2017)

(continued)

Table 4.1 (continued)

Nanomaterials	Recognition element	Target element	Sensing technique	Application	References
Carbon nanotubes	Enzyme	ATP	Optical	Detection of cellular ATP in living cells	Kim et al. (2010)
Graphene	Oligonucleotides	*Escherichia coli* O157:H7	Electrochemical	Detection of *Escherichia coli*	Xu et al. (2017)
Graphene	Enzyme	Bisphenol A (BPA)	Electrochemical	Detection of estrogenic substrate	Reza et al. (2015)
Graphene	Aptamer	Cytochrome c (Cyt c) and caspase-3	Fluorometric	To study cascade reaction in apoptotic signaling	Liu et al. (2018a)
Graphene	Oligonucleotide	Chymotrypsin	Fluorometric	To detect pancreatic diseases	Li et al. (2017)
Quantum dots	Antibody	N-terminal pro-B-type natriuretic peptide (BNP)	Lateral flow immunoassay	Detection of blood biomarker for diagnosing cardiac distress	Wilkins et al. (2018)
Quantum dots	Enzyme	Glucose	Fluorometric	Detection of glucose levels in human biological fluids	Vaishanav et al. (2017)
Quantum dots	Aptamer	Acetamiprid	Fluorometric	In situ visual determination for pesticide residues in complex sample system.	Lin et al. (2016)
Quantum dots	Aptamer	Thrombin	Fluorometric	Detection of protein	Lao et al. (2016)
Nanorods	Aptamer	Human protein tyrosine kinase-7 (PTK-7)	Fluorometric and surface enhanced Raman scattering (SERS)	Cervical cancer diagnostic	Bamrungsap et al. (2016)
Nanorods	Antibody	Activated leukocyte cell adhesion molecule (ALCAM)	Localized surface plasmon resonance (LSPR)	Cancer diagnostic	Pai et al. (2017)

Nanowires	Enzyme	Glucose	Electrochemical	Glucose detection in human serum sample	Li et al. (2015)
Nanowires	Oligonucleotides	Hepatitis B virus (HBV)	Electrochemical	Detection of Hepatitis B virus disease	Shariati (2018)

alternative of the blood sample, which can prevent the painful blood collection and infection risks. They projected a method to couple Fe/Pd/rGO with a portable paper sensor for urine glucose detection. From this method urine glucose was sensitively detected in a broad range of 1–200 μM with a limit of detection of 1.76 μM. The color difference with increasing concentration of urine glucose was easily visualized by naked eyes, which is significant to its realistic usage in screening and diagnosis of diabetes in early-stage (Yang et al. 2019).

Song et al. developed a novel strategy for clinical detection of Zn^{2+} in the human body based on the "on–off–on" ratiometric fluorescent nanosensor using the coupled quantum dots-carbon dots (Song et al. 2018). Copper is directly associated with liver damage; therefore, it is necessary to develop a simple and sensitive strategy to detect copper ions (Cu^{2+}) in liver cells.

Lu et al. developed hydrophobic carbon dots (HCDs)-based dual-emission fluorescent probe for Cu^{2+} detection. The developed probe showed high sensitivity and selectivity to Cu^{2+} above former substances, and it was used to determine the changes of Cu^{2+} level in liver cells (Lu et al. 2017). Qian et al. developed a QDs nanosensor for the detection of 2,4,6-trinitrotoluene (TNT) explosives concentration over a range of 10 nM to 8 μM with a low detection limit of 3.3 nM. One can perform onsite visual determination of TNT with a high resolution because the ratiometric fluorescence nanosensing system exhibited visible fluorescence color changes (Qian et al. 2016).

4.4.2 Detection of Biomolecular Interaction

The elevated sensitivity of the localized surface plasmon resonance (LSPR) spectrum of nanomaterials to adsorbate induced changes in the local refractive index is being used to develop a different class of nanobiosensors or chemosensor (Baida et al. 2009; Brockman et al. 2000; Huang et al. 2009; Nguyen et al. 2015).The optical biosensor detects changes in local refractive index prototypical immunoassay involving biotin (B) and anti-biotin (AB) by monitoring the LSPR (Riboh et al. 2003). Liu et al. (2018b) studied an improved biomolecular interaction analysis probing with an incorporated array fluorescent biosensor with a wide range of low to high concentrations of anti-bisphenol A (BPA) antibody. This scheme is characterized by its high-throughput cross-reactivity analysis between antigens and antibodies on the patterned waveguide with a shortened time-to-result (Liu et al. 2018b). Recently, Zhang et al. prepared biosensor by Ag nanocubes/chitosan composite for the detection of mouse IgG and used to amplify the SPR signal (Zhang et al. 2016). A direct reasonable enzyme-linked immunosorbent assay format was developed and optimized on the surface of a carbon electrode by immobilizing the antibody using an electro-deposition of gold nanoparticles conjugated with polyclonal anti-tetracycline antibodies for detection of tungro disease in paddy plantations (Uda et al. 2019).

4.4.3 Pathogenic Bacteria Detection

Traditionally, bacteria identification in culture and biochemical testing are based on morphology. However, these methods are tiresome and occasionally all bacteria do not grow in culture and therefore the method of detection of bacteria in clinical samples needs to be developed. Magnetic nanoparticles have been used for the identification of *Mycobacterium avium* spp. *paratuberculosis* (MAP) through magnetic relaxation by Kaittanis et al. (2007). Nanobiosensor can be used for direct detection of the pathogenic agent while indirect detection is achievable by evaluating the pathogen's metabolic activity by monitoring the nutrient utilization rate in solution. The dextran-coated gold nanoparticle-based technique is used for the measurement of antimicrobial susceptibility (Nath et al. 2008). *Salmonella* was detected by a gold/silicon nanorods immobilized with dye molecules (Fu et al. 2008). The dye molecules immobilized with silicon nanorods produced fluorescence when it comes in contact with *Salmonella*. This method has remarkable potential in biomedical diagnostics.

Some researcher has also developed a protocol for a simpler diagnostic technique for bacteria. These techniques have unique features that combine magnetic and fluorescent parameters in a nanoparticle-based platform. The magneto-fluorescent nanosensor (MFnS) has been developed that detected *E. coli* O157:H7 contamination with very high sensitivity, 1 colony-forming unit present in water can be detected (Banerjee et al. 2016; Shelby et al. 2017). This method has also been used to detect and quantify the pathogen contamination in both early- and late-stage contamination (Song et al. 2016).

4.4.4 Application in Cancer Biology

Early cancer detection and cure is a rising and attractive ground for research in plasmonic nanobiosensor. Telomerase is a specific reverse transcriptase enzyme containing catalytic subunit and RNA component that maintains the length and function of the telomer together with the proteins associated with the telomere (Greider and Blackburn 1985; Van Steensel and De Lange 1997; Van Steensel et al. 1998). Eventually, a critical telomere length is reached in normal cells, inducing cellular senescence and ultimately leading to apoptosis. In most malignancies, high levels of telomerase activity are found and are thought to play an important role in tumorigenesis (Bodnar et al. 1998; Counter et al. 1998; Kim et al. 1994; Shay and Bacchetti 1997). Telomere dysfunction also causes genetic instability with complex cellular and molecular responses involving checkpoints and apoptosis pathways for the retinoblastoma gene/p53 (Lan et al. 2003; Leri et al. 2003). Grimm et al have developed a magnetic-based nanobiosensor for fast screening of telomerase activity in biological samples (Gullberg et al. 2004). The telomerase-synthesized telomeric repeats (TTAGGG) annealed upon nanoparticles to change their magnetic state (a phenomenon readily detectable by magnetic readers), the developed magnetic nanosensor can determine telomerase activity.

Recently, an optical fiber nanobiosensor was built to efficiently sense a broad-spectrum cancer biomarker, telomerase with its nanoscale tip at a single cell level (Zheng and Li 2010).

Injectable biosensors might be providing a new concept for prostate cancer biomarkers by querying the status of the prostate via a non-invasive readout. Proteases enzyme plays an important role in every characteristic of cancer; its activities could act as biomarkers. Dudani et al. developed a nanosensor library by a panel of prostate cancer proteases through transcriptomic and proteomic analysis that detect protease activity in vitro and in vivo using fluorescence and urinary readouts (Dudani et al. 2018). The impact of nanomaterials on the accuracy of biosensors in early detection of cancer such as lung, prostate, breast, and other cancers was discussed and reviewed by Sharifi et al. The modification of electrode performance by nanomaterials, however, is relatively complicated, resulting in limitations on the use of certain nanomaterials in biosensor applications (Sharifi et al. 2018).Wang et al. developed a small device that facilitated fast and straight analysis of the specific binding of small molecules to proteins using silicon nanowire (SiNW) field-effect transistor (FET) devices (Wang et al. 2005). Chandra et al. designed a sensor probe made-up by immobilizing monoclonal permeability glyco-protein antibody on the gold nanoparticles (AuNPs) conducting polymer composite to detect cancer cells between 50 and 100,000 cells/ml with the detection limit of 23 ± 2 cells/ml (Chandra et al. 2015). Thus, nanosensor can help in the diagnosis of fatal diseases like cancer at an early stage.

4.5 Conclusions and Future Prospects

There is a great demand of analytical devices for rapid, reliable, and economical detection of substances in biological fluids. These devices will be commercialized only if they can be used by a common man rather than by centralized labs or doctor's clinics; biosensors are competent enough to solve these issues. Although biosensors have applications in various fields, their use in health care monitoring is the most important. Nanobiosensor is probed containing immobilized biological molecules on the nanomaterial. They can be used in the detection of microorganisms, pollutants, molecular biomarkers, and monitoring of metabolites in body fluids. Different kinds of nanomaterials like metal, QD, graphene, carbon nanotubes, etc., can be used in the biosensor, each having its own characteristic property. For example, the SPR property of metal nanoparticles is exploited in optical biosensors. Most of the biosensors which are commercialized are enzyme-based sensor as enzymes are highly specific biomolecules. Developments of fabrication methods for nanomaterials which are nontoxic and economically viable are required. Research is required to improve the properties of the nanobiosensor so that it can detect and quantify biological fluids without multiple calibrations using clinical samples. Efforts should be made to improve the sensitivity of detection with high sensitivity. Thus, nanobiosensor should be low cost, disposable, reliable, and easy to use that can be utilized for in-home medical diagnosis of diseases.

Acknowledgements The author is thankful to the University Grant Commission (UGC) for providing financial support.

References

Ali ME, Hashim U, Mustafa S, Che Man YB, Yusop MH, Kashif M, Dhahi TS, Bari MF, Hakim MA, Latif MA (2011) Nanobiosensor for detection and quantification of DNA sequences in degraded mixed meats. J Nanomater 2011:32–43

Ambhorkar P, Wang Z, Ko H, Lee S, Koo KI, Kim K, Cho DI (2018) Nanowire-based biosensors: from growth to applications. Micromachines 9:679

Androvitsaneas P, Young AB, Schneider C, Maier S, Kamp M, Höfling S, Knauer S, Harbord E, Hu CY, Rarity JG, Oulton R (2016) Charged quantum dot micropillar system for deterministic light-matter interactions. Phys Rev B 93:241409

Appenzeller J, Martel R, Derycke V, Radosavljević M, Wind S, Neumayer D, Avouris P (2002) Carbon nanotubes as potential building blocks for future nanoelectronics. Microelectron Eng 64:391–397

Asal M, Ozen O, Şahinler M, Polatoğlu İ (2018) Recent developments in enzyme, DNA and immuno-based biosensors. Sensors 18:1924

Bahadır EB, Sezgintürk MK (2015) Applications of electrochemical immunosensors for early clinical diagnostics. Talanta 132:162–174

Baida H, Billaud P, Marhaba S, Christofilos D, Cottancin E, Crut A, Lermé J, Maioli P, Pellarin M, Broyer M, Del Fatti N (2009) Quantitative determination of the size dependence of surface plasmon resonance damping in single Ag@ SiO_2 nanoparticles. Nano Lett 9:3463–3469

Bakalova R, Ohba H, Zhelev Z, Ishikawa M, Baba Y (2004) Quantum dots as photosensitizers? Nat Biotechnol 22:1360–1361

Bamrungsap S, Treetong A, Apiwat C, Wuttikhun T, Dharakul T (2016) SERS-fluorescence dual mode nanotags for cervical cancer detection using aptamers conjugated to gold-silver nanorods. Microchim Acta 183:249–256

Banerjee T, Sulthana S, Shelby T, Heckert B, Jewell J, Woody K, Karimnia V, McAfee J, Santra S (2016) Multiparametric magneto-fluorescent nanosensors for the ultrasensitive detection of Escherichia coli O157: H7. ACS Infect Dis 2:667–673

Barahona F, Bardliving CL, Phifer A, Bruno JG, Batt CA (2013) An aptasensor based on polymer-gold nanoparticle composite microspheres for the detection of malathion using surface-enhanced Raman spectroscopy. Ind Biotechnol 9:42–50

Baruah S, Dutta J (2009) Nanotechnology applications in pollution sensing and degradation in agriculture: a review. Environ Chem Lett 7:191–204

Bertók T, Katrlík J, Gemeiner P, Tkac J (2013) Electrochemical lectin based biosensors as a label-free tool in glycomics. Microchim Acta 180:1–13

Bhushan B (2007) Nanotribology and nanomechanics of MEMS/NEMS and BioMEMS/BioNEMS materials and devices. Microelectron Eng 84:387–412

Bodnar AG, Ouellette M, Frolkis M, Holt SE, Chiu CP, Morin GB, Harley CB, Shay JW, Lichtsteiner S, Wright WE (1998) Extension of life-span by introduction of telomerase into normal human cells. Science 279:349–352

Borisov SM, Wolfbeis OS (2008) Optical biosensors. Chem Rev 108:423–461

Borman S (1987) Optical and piezoelectric biosensors. Anal Chem 59:1161A–1164A

Brockman JM, Nelson BP, Corn RM (2000) Surface plasmon resonance imaging measurements of ultrathin organic films. Annu Rev Physiol 51:41–63

Bulovic V, Mandell A, Perlman A (2004) Molecular memory device. U.S. Patent No. 6,781,868. Washington, DC: U.S. Patent and Trademark Office

Cai H, Xu C, He P, Fang Y (2001) Colloid Au-enhanced DNA immobilization for the electrochemical detection of sequence-specific DNA. J Electroanal Chem 510:78–85

Cao Z, Zeng M-X, Zhang L, Huang X-X, Wang M-X, Gong F-C, Tan S-Z (2009) Novel optical nanobiosensor assembled with silver nanoparticles on gold surface. In: 2009 4th IEEE international conference on nano/micro engineered and molecular systems. IEEE, Piscataway, pp 987–990

Cao K, Jiang X, Yan S, Zhang L, Wu W (2014) Phenylboronic acid modified silver nanoparticles for colorimetric dynamic analysis of glucose. Biosens Bioelectron 52:188–195

Cash KJ, Clark HA (2010) Nanosensors and nanomaterials for monitoring glucose in diabetes. Trends Mol Med 16:584–593

Cavalcanti A, Shirinzadeh B, Freitas R Jr, Hogg T (2007) Nanorobot architecture for medical target identification. Nanotechnology 19:015103

Cavalcanti A, Shirinzadeh B, Kretly LC (2008) Medical nanorobotics for diabetes control. Nanomed Nanotechnol 4:127–138

Chandra P, Noh H-B, Pallela R, Shim Y-B (2015) Ultrasensitive detection of drug resistant cancer cells in biological matrixes using an amperometric nanobiosensor. Biosens Bioelectron 70:418–425

Chaubey A, Malhotra B (2002) Mediated biosensors. Biosens Bioelectron 17:441–456

Chen S, Hai X, Chen X-W, Wang J-H (2014) In situ growth of silver nanoparticles on graphene quantum dots for ultrasensitive colorimetric detection of H_2O_2 and glucose. Anal Chem 86:6689–6694

Chen C, Yuan Z, Chang H-T, Lu F, Li Z, Lu C (2016) Silver nanoclusters as fluorescent nanosensors for selective and sensitive nitrite detection. Anal Methods 8:2628–2633

Chuang Y-H, Chang Y-T, Liu K-L, Chang H-Y, Yew T-R (2011) Electrical impedimetric biosensors for liver function detection. Biosens Bioelectron 28:368–372

Counter CM, Meyerson M, Eaton EN, Ellisen LW, Caddle SD, Haber DA, Weinberg RA (1998) Telomerase activity is restored in human cells by ectopic expression of hTERT (hEST2), the catalytic subunit of telomerase. Oncogene 16:1217–1222

Cruz HJ, Rosa CC, Oliva AG (2002) Immunosensors for diagnostic applications. Parasitol Res 88: S4–S7

D'souza S (2001) Microbial biosensors. Biosens Bioelectron 16:337–353

Darestani-Farahani M, Faridbod F, Ganjali MR (2018) A sensitive fluorometric DNA nanobiosensor based on a new fluorophore for tumor suppressor gene detection. Talanta 190:140–146

Dasary SS, Rai US, Yu H, Anjaneyulu Y, Dubey M, Ray PC (2008) Gold nanoparticle based surface enhanced fluorescence for detection of organophosphorus agents. Chem Phys Lett 460:187–190

del Valle M, Bonanni A (2014) Impedimetric DNA biosensors based on nanomaterials. Biosens Nanotechnol 18:81–110

Devi R, Yadav S, Nehra R, Yadav S, Pundir C (2013) Electrochemical biosensor based on gold coated iron nanoparticles/chitosan composite bound xanthine oxidase for detection of xanthine in fish meat. J Food Eng 115:207–214

Di Giusto DA, Wlassoff WA, Gooding JJ, Messerle BA, King GC (2005) Proximity extension of circular DNA aptamers with real-time protein detection. Nucleic Acids Res 33:e64–e64

Dudani JS, Ibrahim M, Kirkpatrick J, Warren AD, Bhatia SN (2018) Classification of prostate cancer using a protease activity nanosensor library. Proc Natl Acad Sci U S A 115:8954–8959

Durán N, Marcato PD (2013) Nanobiotechnology perspectives. Role of nanotechnology in the food industry: a review. Int J Food Sci Technol 48:1127–1134

Durán N, Marcato PD, Conti RD, Alves OL, Costa F, Brocchi M (2010) Potential use of silver nanoparticles on pathogenic bacteria, their toxicity and possible mechanisms of action. J Braz Chem Soc 21:949–959

Durmuş NG, Lin RL, Kozberg M, Dermici D, Khademhosseini A, Demirci U (2015) Acoustic-based biosensors. In: Encyclopedia of microfluidics and nanofluidics. Springer, New York, pp 8–40

Dzyadevych SV, Arkhypova VN, Martelet C, Jaffrezic-Renault N, Chovelon JM, El'skaya AV, Soldatkin AP (2004) Potentiometric biosensors based on ISFETs and immobilized cholinesterases. Electroanalysis 16:1873–1882

Eissa S, Alshehri N, Rahman AMA, Dasouki M, Abu-Salah KM, Zourob M (2018) Electrochemical immuno sensors for the detection of survival motor neuron (SMN) protein using different carbon nanomaterials-modified electrodes. Biosens Bioelectron 101:282–289

Fan X, White IM, Shopova SI, Zhu H, Suter JD, Sun Y (2008) Sensitive optical biosensors for unlabeled targets: a review. Anal Chim Acta 620:8–26

Feng X, Gan N, Zhang H, Li T, Cao Y, Hu F, Jiang Q (2016) Ratiometric biosensor array for multiplexed detection of microRNAs based on electrochemiluminescence coupled with cyclic voltammetry. Biosens Bioelectron 75:308–314

Frasconi M, Tel-Vered R, Riskin M, Willner I (2010) Surface plasmon resonance analysis of antibiotics using imprinted boronic acid-functionalized Au nanoparticle composites. Anal Chem 82:2512–2519

Freitas RA (1999) Nanomedicine, volume I: basic capabilities vol 1. Landes Bioscience, Georgetown

Fu J, Park B, Siragusa G, Jones L, Tripp R, Zhao Y, Cho Y-J (2008) An Au/Si hetero-nanorod-based biosensor for Salmonella detection. Nanotechnology 19:155502

Gäberlein S, Spener F, Zaborosch C (2000) Microbial and cytoplasmic membrane-based potentiometric biosensors for direct determination of organophosphorus insecticides. Appl Microbiol Biotechnol 54:652–658

Gomes RC, Abib Pastore VA, Martins OA, Biondi GF (2015) Nanotechnology applications in the food industry: a review. Braz J Hyg Anim Sanity 1:1–8

Gordijo CR, Koulajian K, Shuhendler AJ, Bonifacio LD, Huang HY, Chiang S, Ozin GA, Giacca A, Wu XY (2011) Nanotechnology-enabled closed loop insulin delivery device: in vitro and in vivo evaluation of glucose-regulated insulin release for diabetes control. Adv Funct Mater 21:73–82

Greider CW, Blackburn EH (1985) Identification of a specific telomere terminal transferase activity in Tetrahymena extracts. Cell 43:405–413

Gullberg M, Gústafsdóttir SM, Schallmeiner E, Jarvius J, Bjarnegård M, Betsholtz C, Landegren U, Fredriksson S (2004) Cytokine detection by antibody-based proximity ligation. Proc Natl Acad Sci U S A 101:8420–8424

Haes AJ, Duyne RPV (2004) Preliminary studies and potential applications of localized surface plasmon resonance spectroscopy in medical diagnostics. Expert Rev Mol Diagn 4:527–537

Haes AJ, Van Duyne RP (2002) A nanoscale optical biosensor: sensitivity and selectivity of an approach based on the localized surface plasmon resonance spectroscopy of triangular silver nanoparticles. J Am Chem Soc 124:10596–10604

He W, Huang CZ, Li YF, Xie JP, Yang RG, Zhou PF, Wang J (2008) One-step label-free optical genosensing system for sequence-specific DNA related to the human immunodeficiency virus based on the measurements of light scattering signals of gold nanorods. Anal Chem 80:8424–8430

Hernández-Ibáñez N, García-Cruz L, Montiel V, Foster CW, Banks CE, Iniesta J (2016) Electrochemical lactate biosensor based upon chitosan/carbon nanotubes modified screen-printed graphite electrodes for the determination of lactate in embryonic cell cultures. Biosens Bioelectron 77:1168–1174

Hu L, Deng L, Alsaiari S, Zhang D, Khashab NM (2014) "Light-on" sensing of antioxidants using gold nanoclusters. Anal Chem 86:4989–4994

Huang Y, Bell MC, Suni II (2008) Impedance biosensor for peanut protein AraH1. Anal Chem 80:9157–9161

Huang C, Bonroy K, Reekman G, Verstreken K, Lagae L, Borghs G (2009) An on-chip localized surface plasmon resonance-based biosensor for label-free monitoring of antigen–antibody reaction. Microelectron Eng 86:2437–2441

Jia K, Bijeon JL, Adam PM, Ionescu RE (2012) Sensitive localized surface plasmon resonance multiplexing protocols. Anal Chem 84:8020–8027

Kaittanis C, Naser SA, Perez JM (2007) One-step, nanoparticle-mediated bacterial detection with magnetic relaxation. Nano Lett 7:380–383

Kelly KL, Coronado E, Zhao LL, Schatz GC (2003) The optical properties of metal nanoparticles: the influence of size, shape, and dielectric environment. J Phys Chem B 107(3):668–677

Kim NW, Piatyszek MA, Prowse KR, Harley CB, West MD, Ho PD, Coviello GM, Wright WE, Weinrich SL, Shay JW (1994) Specific association of human telomerase activity with immortal cells and cancer. Science 266:2011–2015

Kim F, Song JH, Yang P (2002) Photochemical synthesis of gold nanorods. J Am Chem Soc 124:14316–14317

Kim JH, Ahn JH, Barone PW, Jin H, Zhang J, Heller DA, Strano MS (2010) A luciferase/single-walled carbon nanotube conjugate for near-infrared fluorescent detection of cellular ATP. Angew Chem Int Ed 49:1456–1459

Ko S, Park TJ, Kim H-S, Kim J-H, Cho Y-J (2009) Directed self-assembly of gold binding polypeptide-protein A fusion proteins for development of gold nanoparticle-based SPR immunosensors. Biosens Bioelectron 24:2592–2597

Koncki R, Radomska A, Głąb S (2000) Potentiometric determination of dialysate urea nitrogen. Talanta 52:13–17

Kueng A, Kranz C, Mizaikoff B (2004) Amperometric ATP biosensor based on polymer entrapped enzymes. Biosens Bioelectron 19:1301–1307

Kulesza A, Mitrić R, Bonačić-Koutecký V (2009) Silver cluster induced absorption enhancement and conformation control of peptides. Eur Phys J D 52:203–206

Kurbanoglu S, Rivas L, Ozkan SA, Merkoçi A (2017) Electrochemically reduced graphene and iridium oxide nanoparticles for inhibition-based angiotensin-converting enzyme inhibitor detection. Biosens Bioelectron 88:122–129

Lad U, Khokhar S, Kale GM (2008) Electrochemical creatinine biosensors. Anal Chem 80:7910–7917

Lan J, Xiong Y-Y, Lin Y-X, Wang B-C, Gong L-L, Xu H-S, Guo G-S (2003) Helicobacter pylori infection generated gastric cancer through p53-Rb tumor-suppressor system mutation and telomerase reactivation. World J Gastroenterol 9:54–58

Landry MP, Ando H, Chen AY, Cao J, Kottadiel VI, Chio L, Yang D, Dong J, Lu TK, Strano MS (2017) Single-molecule detection of protein efflux from microorganisms using fluorescent single-walled carbon nanotube sensor arrays. Nat Nanotechnol 12:368–377

Lao Y-H, Chi C-W, Friedrich SM, Peck K, Wang T-H, Leong KW, Chen L-C (2016) Signal-on protein detection via dye translocation between aptamer and quantum dot. ACS Appl Mater Interfaces 8:12048–12055

Lavan DA, McGuire T, Langer R (2003) Small-scale systems for in vivo drug delivery. Nat Biotechnol 21:1184–1191

Leary SP, Liu CY, Apuzzo ML (2006) Toward the emergence of nanoneurosurgery: part III--nanomedicine: targeted nanotherapy, nanosurgery, and progress toward the realization of nanoneurosurgery. Neurosurgery 58:1009–1026

Leri A, Franco S, Zacheo A, Barlucchi L, Chimenti S, Limana F, Nadal-Ginard B, Kajstura J, Anversa P, Blasco MA (2003) Ablation of telomerase and telomere loss leads to cardiac dilatation and heart failure associated with p53 upregulation. EMBO J 22:131–139

Li X, Zhao C, Liu X (2015) A paper-based microfluidic biosensor integrating zinc oxide nanowires for electrochemical glucose detection. Microsyst Nanoeng 1:15014

Li S, Fu Y, Ma X, Zhang Y (2017) Label-free fluorometric detection of chymotrypsin activity using graphene oxide/nucleic-acid-stabilized silver nanoclusters hybrid materials. Biosens Bioelectron 88:210–216

Liao QG, Wei BH, Luo LG (2017) Aptamer based fluorometric determination of kanamycin using double-stranded DNA and carbon nanotubes. Microchim Acta 184:627–632

Liedberg B, Nylander C, Lunström I (1983) Surface plasmon resonance for gas detection and biosensing. Sensors Actuators 4:299–304
Lin Y, Lu F, Tu Y, Ren Z (2004) Glucose biosensors based on carbon nanotube nanoelectrode ensembles. Nano Lett 4:191–195
Lin K-C, Hong C-P, Chen S-M (2013) Simultaneous determination for toxic ractopamine and salbutamol in pork sample using hybrid carbon nanotubes. Sensors Actuators B Chem 177:428–436
Lin B, Yu Y, Li R, Cao Y, Guo M (2016) Turn-on sensor for quantification and imaging of acetamiprid residues based on quantum dots functionalized with aptamer. Sensors Actuators B Chem 229:100–109
Link S, Wang ZL, El-Sayed M (1999) Alloy formation of gold– silver nanoparticles and the dependence of the plasmon absorption on their composition. J Phys Chem B 103:3529–3533
Liu J-Q, Shimohara K (2007) Molecular computation and evolutionary wetware: a cutting-edge technology for artificial life and nanobiotechnologies. IEEE Trans Syst 37:325–336
Liu Y, Yan J, Howland MC, Kwa T, Revzin A (2011) Micropatterned aptasensors for continuous monitoring of cytokine release from human leukocytes. Anal Chem 83:8286–8292
Liu C, Hu Y-L, Deng W-J, Pan Q-S, Yi J-T, Chen T-T, Chu X (2018a) A graphene oxide nanosensor enables the co-delivery of aptamer and peptide probes for fluorescence imaging of a cascade reaction in apoptotic signaling. Analyst 143:208–214
Liu L, Zhou X, Ma R, He M, Shi H, Yi Q (2018b) High-throughput biomolecular interaction analysis probing by an array fluorescent biosensor platform. Sensors Actuators B Chem 259:888–893
Lu L, Feng C, Xu J, Wang F, Yu H, Xu Z, Zhang W (2017) Hydrophobic-carbon-dot-based dual-emission micelle for ratiometric fluorescence biosensing and imaging of Cu^{2+} in liver cells. Biosens Bioelectron 92:101–108
Lv Y et al (2009) Silver nanoparticle-decorated porous ceramic composite for water treatment. J Membr Sci 331:50–56
Marcato PD, Durán N (2008) New aspects of nanopharmaceutical delivery systems. J Nanosci Nanotechnol 8:2216–2229
Mikkelsen SR, Rechnitz GA (1989) Conductometric transducers for enzyme-based biosensors. Anal Chem 61:1737–1742
Muhammad-Tahir Z, Alocilja EC (2003) A conductometric biosensor for biosecurity. Biosens Bioelectron 18:813–819
Nath S, Kaittanis C, Tinkham A, Perez JM (2008) Dextran-coated gold nanoparticles for the assessment of antimicrobial susceptibility. Anal Chem 80:1033–1038
Nguyen T, Hilton JP, Lin Q (2009) Emerging applications of aptamers to micro-and nanoscale biosensing. Microfluid Nanofluid 6:347
Nguyen H, Park J, Kang S, Kim M (2015) Surface plasmon resonance: a versatile technique for biosensor applications. Sensors 15:10481–10510
Odenthal KJ, Gooding JJ (2007) An introduction to electrochemical DNA biosensors. Analyst 132:603–610
Pai J-H, Yang C-T, Hsu H-Y, Wedding AB, Thierry B (2017) Development of a simplified approach for the fabrication of localised surface plasmon resonance sensors based on gold nanorods functionalized using mixed polyethylene glycol layers. Anal Chim Acta 974:87–92
Pak SC, Penrose W, Hesketh PJ (2001) An ultrathin platinum film sensor to measure biomolecular binding. Biosens Bioelectron 16:371–379
Pandikumar A, How GT, See TP, Omar FS, Jayabal S, Kamali KZ, Yusoff N, Jamil A, Ramaraj R, John SA, Lim HN (2014) Graphene and its nanocomposite material based electrochemical sensor platform for dopamine. RSC Adv 4:63296–63323
Pandit S, Dasgupta D, Dewan N, Prince A (2016) Nanotechnology based biosensors and its application. Pharm Innov 5:18

Pang Y, Wang C, Wang J, Sun Z, Xiao R, Wang S (2016) Fe_3O_4@ Ag magnetic nanoparticles for microRNA capture and duplex-specific nuclease signal amplification based SERS detection in cancer cells. Biosens Bioelectron 79:574–580

Panpradist N, Lai JJ (2016) Point-of-care diagnostics. In: Biomaterials nanoarchitectonics. Elsevier, London, pp 139–156

Park K, Drummy LF, Wadams RC, Koerner H, Nepal D, Fabris L, Vaia RA (2013) Growth mechanism of gold nanorods. Chem Mater 25:555–563

Patel GM, Patel GC, Patel RB, Patel JK, Patel M (2006) Nanorobot: a versatile tool in nanomedicine. J Drug Target 14:63–67

Pavlov V, Xiao Y, Shlyahovsky B, Willner I (2004) Aptamer-functionalized Au nanoparticles for the amplified optical detection of thrombin. J Am Chem Soc 126:11768–11769

Prasad S (2014) Nanobiosensors: the future for diagnosis of disease? Nanobiosens Dis Diagn 3:1–10

Qian J, Hua M, Wang C, Wang K, Liu Q, Hao N, Wang K (2016) Fabrication of l-cysteine-capped CdTe quantum dots based ratiometric fluorescence nanosensor for onsite visual determination of trace TNT explosive. Anal Chim Acta 946:80–87

Radko S, Rakhmetova SY, Bodoev N, Archakov A (2007) Aptamers as affinity reagents for clinical proteomics. Biochemistry 1:198–209

Rai M, Yadav A, Gade A (2009) Silver nanoparticles as a new generation of antimicrobials. Biotechnol Adv 27:76–83

Rai M, Gade A, Gaikwad S, Marcato PD, Durán N (2012) Biomedical applications of nanobiosensors: the state-of-the-art. J Braz Chem Soc 23:14–24

Razavi H, Janfaza S (2015) Medical nanobiosensors: a tutorial review. Nanomed J 2:74–87

Reza KK, Ali MA, Srivastava S, Agrawal VV, Biradar A (2015) Tyrosinase conjugated reduced graphene oxide based biointerface for bisphenol A sensor. Biosens Bioelectron 74:644–651

Riboh JC, Haes AJ, McFarland AD, Ranjit Yonzon C, Van Duyne RP (2003) A nanoscale optical biosensor: real-time immunoassay in physiological buffer enabled by improved nanoparticle adhesion. J Phys Chem B 107:1772–1780

Ronkainen NJ, Halsall HB, Heineman WR (2010) Electrochemical biosensors. Chem Soc Rev 39:1747–1763

Sahoo SK, Misra R, Parveen S (2017) Nanoparticles: a boon to drug delivery, therapeutics, diagnostics and imaging. Nanomed Cancer 8:73–124

Samuel D, Bharali D, Mousa SA (2010) The role of nanotechnology in diabetes treatment: current and future perspectives. Int J Nanotechnol 8:53–65

Santandreu M, Alegret S, Fabregas E (1999) Determination of β-HCG using amperometric immunosensors based on a conducting immunocomposite. Anal Chim Acta 396:181–188

Sassolas A, Blum LJ, Leca-Bouvier BD (2009) Electrochemical aptasensors. Electroanalysis 21:1237–1250

Sener G, Uzun L, Denizli A (2014) Colorimetric sensor array based on gold nanoparticles and amino acids for identification of toxic metal ions in water. ACS Appl Mater Interfaces 6:18395–18400

Shariati M (2018) The field effect transistor DNA biosensor based on ITO nanowires in label-free hepatitis B virus detecting compatible with CMOS technology. Biosens Bioelectron 105:58–64

Sharifi M, Avadi MR, Attar F, Dashtestani F, Ghorchian H, Rezayat SM, Saboury AA, Falahati M (2018) Cancer diagnosis using nanomaterials based electrochemical nanobiosensors. Biosens Bioelectron 126:773–784

Sharma A, Rogers KR (1994) Biosensors measurement. Sci Technol 5:461

Shay J, Bacchetti S (1997) A survey of telomerase activity in human cancer. Eur J Cancer 33:787–791

Sheehan PE, Whitman LJ (2005) Detection limits for nanoscale biosensors. Nano Lett 5:803–807

Shelby T, Sulthana S, McAfee J, Banerjee T, Santra S (2017) Foodborne pathogen screening using magneto-fluorescent nanosensor: rapid detection of E. coli O157: H7. J Vis Exp 127:e55821

Shi H, Xia T, Nel AE, Yeh JI (2007) Part II: coordinated biosensors–development of enhanced nanobiosensors for biological and medical applications. Nanomedicine 2:599–614

Shi S, Wang X, Sun W, Wang X, Yao T, Ji L (2013) Label-free fluorescent DNA biosensors based on metallointercalators and nanomaterials. Methods 64:305–314

Shin HJ (2011) Genetically engineered microbial biosensors for in situ monitoring of environmental pollution. Appl Microbiol Biotechnol 89:867–877

Shirale DJ, Bangar MA, Park M, Yates MV, Chen W, Myung NV, Mulchandani A (2010) Label-free chemiresistive immunosensors for viruses. Environ Sci Technol 44:9030–9035

Sistani P, Sofimaryo L, Masoudi ZR, Sayad A, Rahimzadeh R, Salehi B (2014) A penicillin biosensor by using silver nanoparticles. Int J Electrochem Sci 9:6201–6212

Song C, Li J, Liu J, Liu Q (2016) Simple sensitive rapid detection of Escherichia coli O157: H7 in food samples by label-free immunofluorescence strip sensor. Talanta 156:42–47

Song Q, Ma Y, Wang X, Tang T, Song Y, Ma Y, Xu G, Wei F, Cen Y, Hu Q (2018) "On-off-on" fluorescent system for detection of Zn^{2+} in biological samples using quantum dots-carbon dots ratiometric nanosensor. J Colloid Interface Sci 516:522–528

Spain E, Gilgunn S, Sharma S, Adamson K, Carthy E, O'Kennedy R, Forster RJ (2016) Detection of prostate specific antigen based on electrocatalytic platinum nanoparticles conjugated to a recombinant scFv antibody. Biosens Bioelectron 77:759–766

Tilmaciu C-M, Morris MC (2015) Carbon nanotube biosensors. Front Chem 3:59

Trojanowicz M (2002) Determination of pesticides using electrochemical enzymatic biosensors. Electroanalysis 14:1311–1328

Turner AP (2000) Biosensors--sense and sensitivity. Science 290:1315–1317

Uda MN, Hasfalina CM, Samsuzanaa AA, Faridah S, Gopinath SC, Parmin NA, Hashim U, Mat M (2019) A disposable biosensor based on antibody-antigen interaction for tungro disease detection. In: Nanobiosensors for biomolecular targeting. Elsevier, London, pp 147–164

Vaishanav SK, Korram J, Nagwanshi R, Ghosh KK, Satnami ML (2017) Mn^{2+} doped-CdTe/ZnS modified fluorescence nanosensor for detection of glucose. Sensors Actuators B Chem 245:196–204

Van Steensel B, De Lange T (1997) Control of telomere length by the human telomeric protein TRF1. Nature 385:740–743

Van Steensel B, Smogorzewska A, De Lange T (1998) TRF2 protects human telomeres from end-to-end fusions. Cell 92:401–413

Velasco-Garcia M (2009) Optical biosensors for probing at the cellular level: a review of recent progress and future prospects. In: Seminars in cell & developmental biology, vol 1. Elsevier, London, pp 27–33

Vo-Dinh T (2005) Optical nanosensors for detecting proteins and biomarkers in individual living cells. In: Protein nanotechnology. Springer, Cham, pp 383–401

Wang J (1999) Amperometric biosensors for clinical and therapeutic drug monitoring: a review. J Pharmaceut Biomed 19:47–53

Wang J, Musameh M (2003) Carbon nanotube/teflon composite electrochemical sensors and biosensors. Anal Chem 75:2075–2079

Wang WU, Chen C, Lin K-H, Fang Y, Lieber CM (2005) Label-free detection of small-molecule–protein interactions by using nanowire nanosensors. Proc Natl Acad Sci U S A 102:3208–3212

Wang Z, Luo X, Wan Q, Wu K, Yang N (2014) Versatile matrix for constructing enzyme-based biosensors. ACS Appl Mater Interfaces 6:17296–17305

Weerathunge P, Ramanathan R, Shukla R, Sharma TK, Bansal V (2014) Aptamer-controlled reversible inhibition of gold nanozyme activity for pesticide sensing. Anal Chem 86:11937–11941

Wilkins MD, Turner BL, Rivera KR, Menegatti S, Daniele M (2018) Quantum dot enabled lateral flow immunoassay for detection of cardiac biomarker NT-proBNP. Sens Biosens Res 21:46–53

Wu S, Duan N, Shi Z, Fang C, Wang Z (2014) Simultaneous aptasensor for multiplex pathogenic bacteria detection based on multicolor upconversion nanoparticles labels. Anal Chem 86:3100–3107

Xu Y, Ding J, Chen H, Zhao Q, Hou J, Yan J, Wang H, Ding L, Ren N (2013) Fast determination of sulfonamides from egg samples using magnetic multiwalled carbon nanotubes as adsorbents followed by liquid chromatography–tandem mass spectrometry. Food Chem 140:83–90

Xu S, Zhang Y, Dong K, Wen J, Zheng C, Zhao S (2017) Electrochemical DNA biosensor based on graphene oxide-chitosan hybrid nanocomposites for detection of Escherichia coli O157: H7. Int J Electrochem Sci 12:3443–3458

Yan J, Guan H, Yu J, Chi D (2013) Acetylcholinesterase biosensor based on assembly of multiwall carbon nanotubes onto liposome bioreactors for detection of organophosphates pesticides. Pestic Biochem Physiol 105:197–202

Yang C, Feng W, Li Y, Tian X, Zhou Z, Lu L, Nie Y (2019) A promising method for diabetes early diagnosis via sensitive detection of urine glucose by FePd/rGO. Dyes Pigments 164:20–26

You C, Bhagawati M, Brecht A, Piehler J (2009) Affinity capturing for targeting proteins into micro and nanostructures. Anal Bioanal Chem 393:1563–1570

Yun YH, Shanov VN, Bange A, Heineman WR, Halsall HB, Seth G, Pixley SK, Behbehani M, Bhattacharya A, Dong Z, Yarmolenko S (2009) Carbon nanotube smart materials for biology and medicine. In: Nanoscience in biomedicine. Springer, New York, pp 451–484

Zhang Z, Lin M, Zhang S, Vardhanabhuti B (2013) Detection of aflatoxin M1 in milk by dynamic light scattering coupled with superparamagnetic beads and gold nanoprobes. J Agric Food Chem 61:4520–4525

Zhang LN, Deng HH, Lin FL, Xu XW, Weng SH, Liu AL, Lin XH, Xia XH, Chen W (2014) In situ growth of porous platinum nanoparticles on graphene oxide for colorimetric detection of cancer cells. Anal Chem 86:2711–2718

Zhang D, Sun Y, Wu Q, Ma P, Zhang H, Wang Y, Song D (2016) Enhancing sensitivity of surface plasmon resonance biosensor by Ag nanocubes/chitosan composite for the detection of mouse IgG. Talanta 146:364–368

Zhao X, Hilliard LR, Mechery SJ, Wang Y, Bagwe RP, Jin S, Tan W (2004) A rapid bioassay for single bacterial cell quantitation using bioconjugated nanoparticles. Proc Natl Acad Sci U S A 101:15027–15032

Zhao W-W, Xu J-J, Chen H-Y (2014) Photoelectrochemical DNA biosensors. Chem Rev 114:7421–7441

Zheng XT, Li CM (2010) Single living cell detection of telomerase over-expression for cancer detection by an optical fiber nanobiosensor. Biosens Bioelectron 25:1548–1552

Nanosensors Based on DNA as an Emerging Technology for the Detection of Disease

5

Akrema, Rizwan Arif, Summaiyya Khan, and Rahisuddin

Abstract

An early age detection of diseases is a big challenge in the healthcare market. There are different conventional methods available for the detection of various diseases, but they are time consuming, labor intensive, require sample preparation as well as need trained personnel. Today, Nanotechnology has potential applications in the field of the public health sector for the detection of viruses and bacteria. Recent advancements in nanosensing technology, especially in the field of biosensors provides a more accurate tool for early diagnosis, reduce cost and fast detection, especially with cardiovascular and diabetes which have high healthcare costs. The DNA-based nanosensors work on different principles which include localized surface plasmon resonance, Raman scattering, fluorescence, and other techniques. The emerging ability of nanoscale materials to control the patterns of matter can lead to entirely new types of nanosensors. The present chapter describes the current status of nanosensors based on DNA in the context of different methods for detection. The conclusion part puts light on the vast possibilities of different types of nanomaterials into sensor arrays and the expected outcomes and limitations.

Keywords

Nanosensors · DNA · Raman · Fluorescence · Surface plasmon resonance

Akrema · R. Arif · S. Khan · Rahisuddin (✉)
Department of Chemistry, Jamia Millia Islamia, New Delhi, India
e-mail: rahisuddin@jmi.ac.in

M. Mohsin et al. (eds.), *Nanobiosensors for Agricultural, Medical and Environmental Applications*, https://doi.org/10.1007/978-981-15-8346-9_5

5.1 Introduction

Nanosensors are nanoscale analytical devices that measure chemical species, nanoparticles, and physical parameters that convert these parameters and quantities into signals for the detection. For the development of nanosensor, it is necessary that sensitivity, spatial confinement of the interaction of the sensor, and size of the nanosensor must be on nanoscale (Liao et al. 2014). Sensors generally convert or transduce information in the form of one form of energy such as thermal, mechanical, electrical, optical, or biochemical into another form. A chemical sensor is very sensitive to detect and analyze a transmitted signal of chemical or biological molecule used. Carbon nanotubes (CNTs), nanowires, and nanoparticles can be used as chemical sensors (Erickson et al. 2008). Nanoscale biosensors base research area is an ideal research area for detection of various diseases in the field of advance nanotechnology and growing rapidly day by day as their high sensitivity, versatility, small size of instrumentation, high specificity, high speed, and low cost which detect the signal of biochemical and biophysical field with a particular disease and do help for detection of microorganisms and monitoring of body fluids for tissue pathology (Dahman et al. 2017; Touhami 2015). Diagrammatic representation for chemical biosensor is shown in Figs. 5.1 and 5.2.

Today, various methods have been proposed for the development of nanosensors including top-down, bottom-up approach, and molecular self-assembly with many

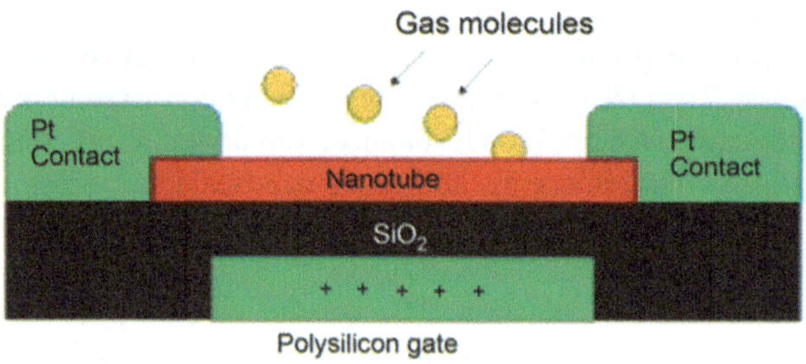

Fig. 5.1 Nanotube-based chemical sensor. Adapted with permission from Erickson et al. (2008)

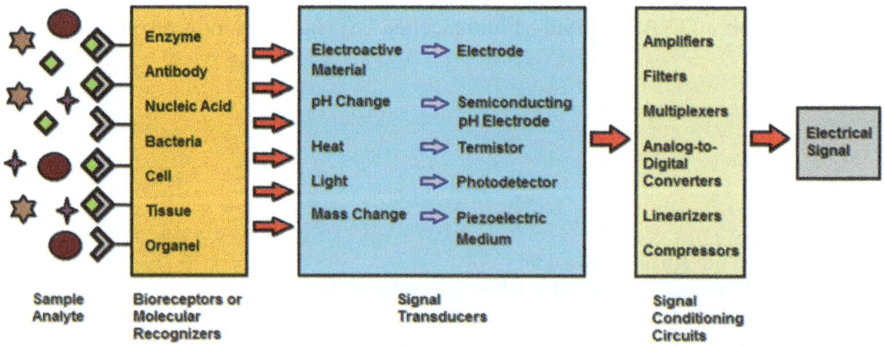

Fig. 5.2 Nanotube-based chemical sensor. Adapted with permission from Erickson et al. (2008)

of the advantages like sensitivity and specificity in comparison to other sensors. An analyte that combines biological component and physicochemical detector for the reorganization of biological recognition element at a nanoscale level has been detected by device denoted as nanobiosensor (Malik et al. 2013). There are many types of nano-bio sensors, i.e., optical, electrochemical, nanotube-based, nanowire, viral, and nanoshell biosensors (Khan et al. 2016). Optical nanobiosensors utilize the light to sense the effect of chemical activity on a biological system and due to the small size of nanosensors sensing of the physiological and biological parameter should be easily done (Yuna et al. 2003). Nanowire based nanobiosensors use as semiconducting device with the fabrication of nanomaterials like carbon nanotubes (CNT), nanowires of metal oxide having greater surface-volume ratio. Piezo-Electric biosensors used piezo-electric devices like quartz, work in an electric field medium. The principle of electrochemical biosensors based on applied current after applying a potential between two electrodes, changes in the charge distribution has been detected through ion-selective electrodes. Due to increase in sensitivity by enhancing the surface-to-volume ratio of nanomaterials and changing their physical properties nanobiosensor operate at the scale of natural biological processes and can be used after the integration of nanosensor with nano-electronics to increase their processing capability which in turn promote the researcher for emerging medical fields and diagnostics (He et al. 2008). Nanobiosensors are used as deoxyribonucleic acid (DNA) and cell-based sensor, as immune-sensors, detection of diseases (HIV, Hepatitis, and viral diseases), genetic and environmental monitoring, drug testing, used for the care of blood, urine, steroids, drugs, hormones, and proteins (Abu-Salah et al. 2015). The bacterial sensor can be used in the food industry, the medicinal field, etc. In cancer monitoring, there is a very important and crucial role of nanobiosensors for body fluids detection. The sensor used for cancer detection was used after the coating of a specific antibody belongs to cancer (Doucey and Carrara 2019).

5.2 DNA-Based Nanobiosensor

Detection of diseases is the very first stage to prevent from any type of disease which is very difficult and many scientists and researchers of various disciplines and subjects are working on this research area for better diagnostics with low cost in limited time. Research area to develop DNA-based nanobiosensor is very interesting for the detection of diseases due to its specificity, sensitivity, and low cost (Rubtsova et al. 2017). In the human body, all the cells contain DNA which carries out genetic information for the physiological activities of human bodies. Every gene of DNA in our body comprises specific genetic code in the form of triplet which gives the instructions for the specific function. Due to its easy implementation, easy control, and low cost, DNA is an ideal material for nanofabrication of nanomaterials for the diagnostic based analyses of pathogenic diseases and genetic disorders (Abu-Salah et al. 2015).

DNA-based nanobiosensor is a device that detects signals and transmits the information on biological analytes like DNA, ribonucleic acid (RNA), and protein with recognition element, transducer, and processor. Magnus and Yi-Ping developed a nanobiosensor for the detection of the functionality of DNA-modifying enzymes, with emphasis on the topoisomerase and tyrosyl-DNA phosphodiesterase families and saw that DNA-nanosensors could be used as clinical diagnostics via enzyme detection. By merging DNA nanosensors and rolling circle amplification is a better idea for the demonstration of specific and sensitive measurements of of DNA topoisomerase and tyrosine recombinase families (Stougaard et al. 2009; Andersen et al. 2009).

Tripathy et al. (2016) synthesized an ultrasensitive electrochemical DNA nanosensor based on semiconducting manganese(III) oxide nanofibers. The adapted electrochemical biosensing technique was used to detect dengue disease. Tripathy and his co-worker found that better sensitivity may be due to the inherent properties of the manganese oxide nanofibers. He also discovered that the electrochemical biosensing technique could also be used for antibody–antigen interactions. A fluorescent DNA nanobiosensor was developed by Darestani-Farahani and co-worker to determine the DNA sequence of the tumor suppressor gene Adenomatous Polyposis Coli (APC) by using gold nanoparticles (AuNPs) for the enhancement of sensitivity. DNA nanobiosensor was developed which was highly sensitive for the evaluation of the APC gene sequence in the real plasma sample (Darestani-Farahania et al. 2018).

Shamsipur et al. (2017) developed a very sensitive and convenient nanobiosensor based on fluorescence resonance energy transfer (FRET) for the detection of a 22-*mer* oligonucleotides sequence in the Human Papillomavirus 18 virus (HPV18) gene. DNA conjugated-quantum dots of CdTe were synthesized with the modification of the amino group in 11-mer oligonucleotides and attached it with the two necessary probes. They prepared a sandwich hybrid of DNA conjugate quantum dots (QDs) and Cyanine5 (Cy5)-labeled 11-mer oligonucleotides. After the demonstration of results, researchers concluded that the interaction intensity of complementary DNA was about 339% and 2826% larger than mismatch and non-complementary sequences and advised that this diagnostic procedure could be further used in the future for specific detection of complementary DNA targets in infected person with various viral pathogens. Biosensors are considered as multifunctional and multicomponent devices due to the reason that they perform various functions at the same time and same platform (Fig. 5.3). It can also be integrated into a wristwatch and cassette-type DNA biosensor (Fig. 5.4) were also developed after the incorporation of microfluidics and advanced biosensor technologies with high-throughput analysis, portability, low reagent consumption, and many favorable properties and several other advantages like minimal handling of hazardous materials, detection of samples in flexibility (Ansari et al. 2016; Bhushan 2010).

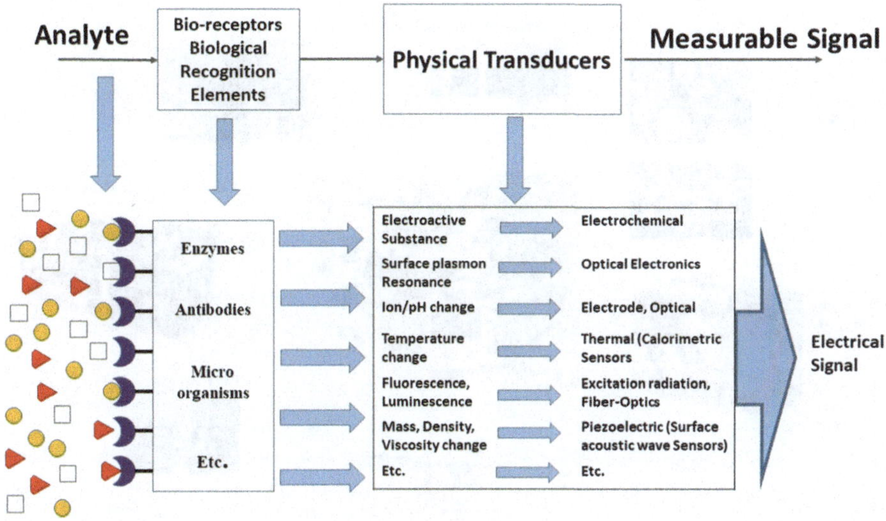

Fig. 5.3 Schematic representation of biosensor and its assembly. Adapted with permission from Ansari et al. (2016)

5.3 Types of Nanobiosensors

5.3.1 Nanostructured Materials

Most of the properties of the solids depend upon its chemical structure. Modification in the size of the solids in one, two, or three dimensions will vary the properties of the solids. The well- known example of this is the change in the electronic optical properties of gold when it changes from bulk to nano. The synthesis process is important in controlling the size of the nanoparticles which has become an emerging interdisciplinary field. There are about three categories of nanostructured materials (Gleiter 1995). The first category includes the materials which have reduced dimensions in the form of thin wires, films, or nanometer sized particles. This type of nanostructured materials can be synthesized using precipitation from the vapor, aerosol techniques, chemical vapor deposition (CVD) and physical vapor deposition (PVD).

The subsequent category includes those materials in which nano-sized structure is limited to a thin surface region of bulk material. This category material can be synthesized using CVD, PVD, laser beam treatments, and ion implantation, which are the most common methods for the synthesis. One more category includes bulk solids with a nan-scale microstructure.

Nanostructured materials are also classified as zero dimension (0D), one dimension (1D), two dimension (2D), and three dimension (3D). 0D includes those materials in which all dimensions are in nanometer scale and includes quantum

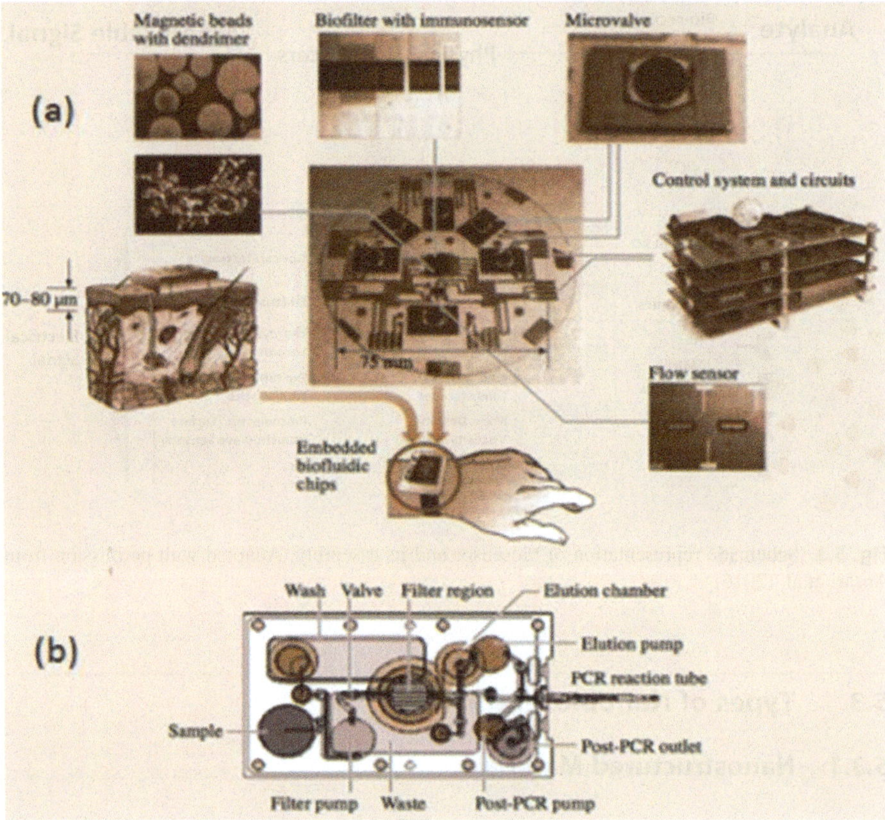

Fig. 5.4 (**a**) Biosensor incorporated wristwatch. The small volume of fluid is manipulated by employing micropumps and microvalves, (**b**) a cassette-type biosensor used for human genomic DNA analysis. The use of micropumps and microvalves allows for fluid manipulation and multiple-unit operations in a single cassette. Adapted with permission from Bhushan (2010)

dots, hollow spheres, nanolenses, etc. 1D nanostructured materials are those in which at least one dimension is in nanoscale and includes nanowires, nanorods, nanotubes, etc. 2D includes materials in which two dimensions are in nanoscale and includes self-assembled monolayer, multilayer, plates, etc. Coming to the last one which is 3D nanostructured materials are those in which no dimensions are in nanoscale. It includes bulk materials, bundles of nanowires, and nanotubes as well as dispersion of nanoparticles.

5.3.2 Nanoparticles

Nanoparticles have the particle size range of 1–100 nm. Presently, intense research is proceeding in the area of nanotechnology. They are of great scientific interest due to their exclusive and specific properties. The physical properties of the bulk material do not change irrespective of its size, but in nanoscale properties change with size. The most important property of nanoparticle is high surface-to-volume ratio which dominates the bulk materials. This means that the proportion of atoms available at the surface of a particle for catalyzing a reaction is inversely proportional to the radius of the particle. Because of this variation in lattice structure at the surface of the nanoparticles, the higher catalytic activity than the bulk material is achieved. Nanoparticles are small enough to produce quantum confinement and therefore, have unique optical properties. The two most common methods for synthesizing nanoparticles are top-down approach and bottom-up approach. Top-down approach implies the cutting of larger particles into smaller ones. This type of synthesis can be done using lithographic techniques, grinding in a ball mill or etching. Bottom-up approach is the more convenient method which involves miniaturization of material components with further self-assembly process leading to the formation of nanoparticles. The bottom-up approach is considered more advantageous compared to the top-down approach because the former produces nanostructures having less defects and more homogeneous chemical composition.

5.3.3 Nanowire/Nanotube

Any solid material having the shape of a wire, with a diameter smaller than 100 nm is referred to as nanowire. There are different types of nanowires such as superconducting, metallic, semiconducting as well as insulating nanowires. Apart from this, molecular nanowires are comprised of recurring molecular arrangements either inorganic or organic. Synthesis of the nanowires can be done using lithography, thermal oxidation, or milling. Several reasons show that nanowires have less conductivity corresponding to the bulk material. A significant effect is seen after the scattering from the nanowire boundaries if the width of the nanowire is less than the electron average free path of the bulk material. The small size of nanowires also contributes towards unique electrical properties. Nanowire conductivity is very much influenced by the edge effect in which the atoms on the surface do not fully bond to neighboring atoms as in the case of the bulk nanowires.

5.3.4 Nanoprobe

Nanoprobe is an optical device. It can also be defined as any chemical or biological technique dealing with nano quantities. They have different applications such as introducing nanoparticles in aqueous solutions that serve as nanoprobes in electrospray ionization mass spectrometry (Hui-Fen et al. 2010) and also in gold-based metallic nanoprobes for therapeutic diagnostics (Balaji et al. 2011).

5.4 Detection of Various Diseases

5.4.1 Diseases Caused by Human Immune Deficiency Virus (HIV)

HIV-1 is caused by a severe disease acquired immune deficiency syndrome (AIDS). According to a report of 2014, "36.9 million people were living with HIV-1, 2 million new cases, and 1.2 million deaths worldwide" (Global AIDS Update 2016). The international occurrence of HIV-1 has declined due to several factors including suppressive antiretroviral therapy and needle replacement programs in developed countries (Abdul-Quader et al. 2013; Mathers et al. 2009; Quinn et al. 2000).

Zheng et al. (2012) proposed a new immune-sensor for HIV p24 by attaching gold nanoparticles on an electrode surface through electroplating by using chrono-amperometry, which increased the conductivity and reversibility of the electrode. After optimization, the relationship between the concentration of p24 and the electrochemical signal was found to be linear, ranging from 0.01 ng ml^{-1} to 100 ng ml^{-1} ($R > 0.99$), and the limit of detection was 0.008 ng ml^{-1}. This observation is better than the conventional ELISA methods by the two orders of magnitude. Researchers developed an immunoassay for HIV protein detection to capture HIV p24 protein, where the electrode was modified with anti-p24 antibodies followed by labeling with horseradish peroxidase (HRP) secondary antibody. The electrodes were immersed in a solution of hydroquinone and hydrogen peroxide which is directly catalyzed by the HRP-secondary antibody complex and the signals were recorded.

In another report, an impedance-based micro device was fabricated to analyze HIV viral nano-lysates by recording impedance changes (Shafiee et al. 2013). Magnetic beads coated with anti-gp120 antibodies and viral lysis were used to capture multiple HIV-1 subtypes. The nano-lysate samples were loaded to micro devices comprising of a Pyrex wafer having two gold microelectrodes. Electro impedance spectroscopy (EIS) data was recorded among multiple HIV-1 subtypes and control from 100 Hz to 1 MHz. The tested samples gave distinct impedance signals compared to control. Epstein-Barr virus-containing HIV-1 samples were also tested and the results were similar to that of HIV-1, which shows the specific detection of HIV. The experimental results show the ability to detect HIV-1 at clinically relevant viral concentrations between 106–109 copies ml^{-1} that appear 3–8 weeks after infection (acute stage). The nano-lysate demonstrated the extraordinary potential to detect viral pathogens directly through the simple impedance signals obtained by an electrical sensing device. The total cost of the reagents and fabrication of the microelectromechanical systems device is less than USD 2, which makes it an affordable tool in the field of label-free electrical detection.

Wang et al. (2015) designed a gold nanocluster modified graphene electrode to measure HIV originated target sequences using an exonuclease III (Exo III) assisted target recycling amplification strategy. authors used Cytosine (C)-rich base capture probe binding to GR/AuNCs platform to bind to gold nanoparticles covered GCE instead of using aptamers modified by –SH or $–NH_2$ as capture probe. GR/AuNCs electrode was loaded with capture probes, which were then labeled with methylene blue on 3′-end and Cytosine (C)-rich base on 5′-end. During hybridization between target and capture probes, nucleic acids folded themselves into a duplex DNA structure followed by the digestion of the capture probe by Exo III from its 3′-end, which results in the release of methylene blue molecules. The changes in the signal were monitored by differential pulse voltammetry and detected down to 30 aM of HIV target probe with a dynamic range 0.1 fM to 100 nM. The recovery of serum samples was reported to be 99.8% at 10 fM concentration of the HIV probe target.

The development of new technologies is based on the detection by fluorescent label and direct count cells, thus eliminating the need for specialized flow cytometer equipment. For example, "Jokerst et al. (2008) have developed QDs based detection method in which labeled CD4 cells with QDs were captured on bio-activated polydimethylsiloxane (PDMS) nano-biochips and imaged using a portable single wavelength fluorescent microscope." CD4+ T cells were distinguished from other monocytes and lymphocytes (Jokerst et al. 2008). Further, the authors designed the membrane-based system to determine CD4 absolute counts in the range of 200–600 ml^{-1}. This method is harmonious with absolute CD4 count, %CD4 determination, CD4/CD8 ratio, and other cellular analyses. On the other side, a research group at Partners AIDS Research Center, Massachusetts General Hospital reported that controlled shear stress can be used in a flow chamber to capture CD4+ T cells and fluorescently labeled antibodies can be used to count the cells (Jokerst et al. 2008). In a simple method, only 10 μL of the whole blood sample was required to count CD4 cell and an antibody functionalized microchannel was used to capture monocytes and CD4+ T cells (Fig. 5.5a, c). The same group of researchers further proposed a two-stage microfluidic device to deplete monocytes to allow CD4+ T cell capture at lower concentrations (Fig. 5.5b, d) (Cheng et al. 2007). The microchip-based device gave comparable CD4 counts with established flow cytometry measurements in the range from 200 μl^{-1} to 800 cells μl^{-1}. At a range lower than 200 μl^{-1} CD4+ monocytes often dominate the cell population in the microchannel which advances to a positive bias. To encounter this problem, researchers came out with a two-stage microfluidic device (Fig. 5.5b).

Caires et al. (2019) have used AuNPs and carboxy methylcellulose (CMC) biopolymer, which acted as a green in-situ reducing and surface stabilizing agent for the water-based colloidal procedure to developed nano-immuno-conjugates. The researchers used AuNPs-CMC nano-colloid with the gp41 glycoprotein receptor or HIV monoclonal antibodies (PolyArg-ab-HIV) for detection. The laser light scattering immunoassay (LIA) was used for detection. The synthesized nano-conjugates formed spherical nanocrystalline AuNPs with size ranging from 12 to 20 nm and surface plasmon resonance at 520 nm. Dynamic light scattering measurements were used to detect the aggregate formation of antibody–antigen nanocomplexes by coupling the immuno-conjugates AuNPs-CMC-gp41 and AuNPs-CMC-PolyArg-ab-HIV. The interference by other proteins, such as bovine serum albumin (BSA), in blood serum was also investigated to get the LIA response. The authors demonstrated a similar trend with and without adding BSA to LIA assay. Thus, the AuNPs-CMC immuno-conjugates at the nanomolar level shows the high binding

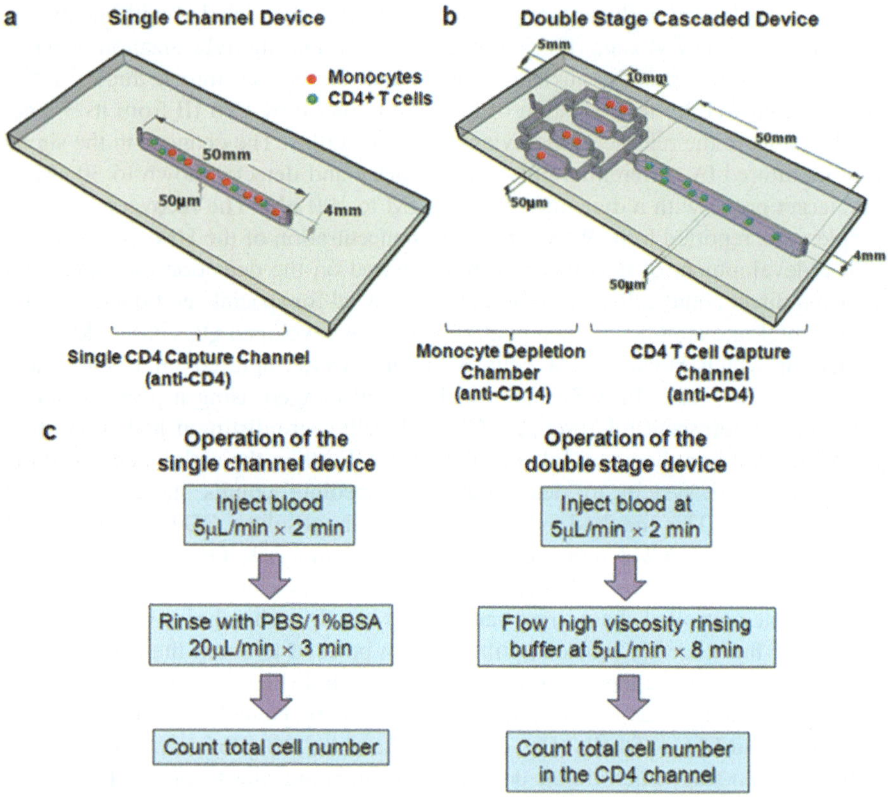

Fig. 5.5 Different types of microfluidic devices. Schematics showing the geometry of the single-channel (**a**) and two-stage (**b**) devices used for cell counting. The operation procedures (**c**) and (**d**) are of two counting devices. Adapted with permission from Jokerst et al. (2008)

affinity, specificity, and sensitivity. For clinical applications, the extraction and purification of the antibody from the patient's blood serum is to be done to conjugate with the AuNPs in order to perform the LIA assay.

Zhou et al. (2013) have used liposome–QD (L-QD) complexes and single-particle detection techniques for multiplexed detection of HIV-1 and HIV-2 DNAs (Fig. 5.6). The reason behind the use of liposomes to encapsulate the QDs was their prominent cargo carrier capability, accessible functionality, and desirable biocompatibility. The research group developed liposome-green and liposome-red QD complexes and designed two sets of oligonucleotide probes for sensitive detection of target DNA. In the first set, target oligonucleotide of HIV-1 was sandwiched between a magnetic bead-modified capture probe 1 and a L-QD green complex-tagged reporter probe 1 on the basis of Watson–Crick base pairing. In the second set, target oligonucleotide of HIV-2 was sandwiched between a magnetic bead-modified capture probe 2 and a L-QD red complex-tagged reporter probe 2. The green and red QDs are released from the liposome–QD complexes after magnetic

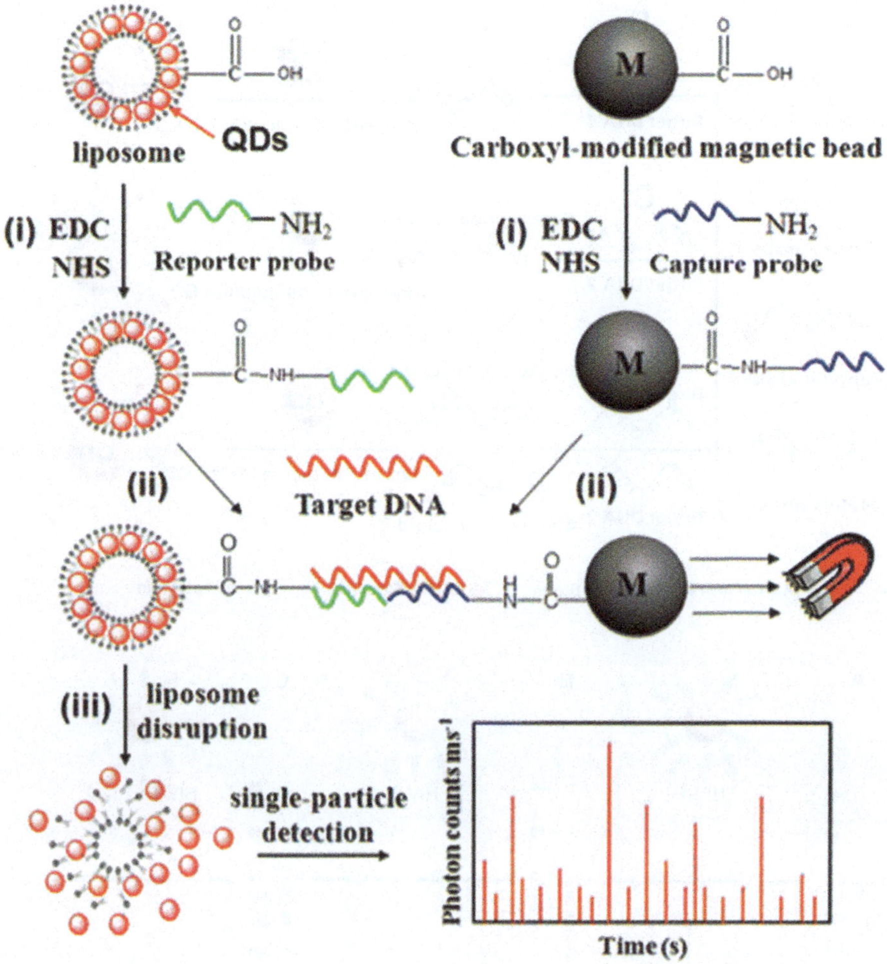

Fig. 5.6 Illustration for detection of attomolar DNA. This method involves three steps: (i) synthesis of L/QD complexes, L/QD complex-tagged reporter probes, and magnetic bead-modified capture probes; (ii) formation of sandwich hybrids; (iii) release of QDs from L/QD complex and measurement by single-particle detection. Adapted with permission from Zhou et al. (2013)

separation and liposome disruption and detected by single QD counting. The results showed a limit of detection up to five orders of magnitude in comparison with fluorescence-tagged and microbead-based sensors (Han et al. 2001) and three orders of magnitude in comparison with single QD-based sensors (Zhang et al. 2005).

In earlier QD-based nanosensor article, Zhang and Hu (2010) used coincidence detection and FRET detection for multiplexed detection of HIV-1 and HIV-2a based on single QD-nanosensor (Fig. 5.7). They reported two sets of two biotinylated capture probes Alexa Fluor 488-labeled reporter probe 1 and capture probe 1 for

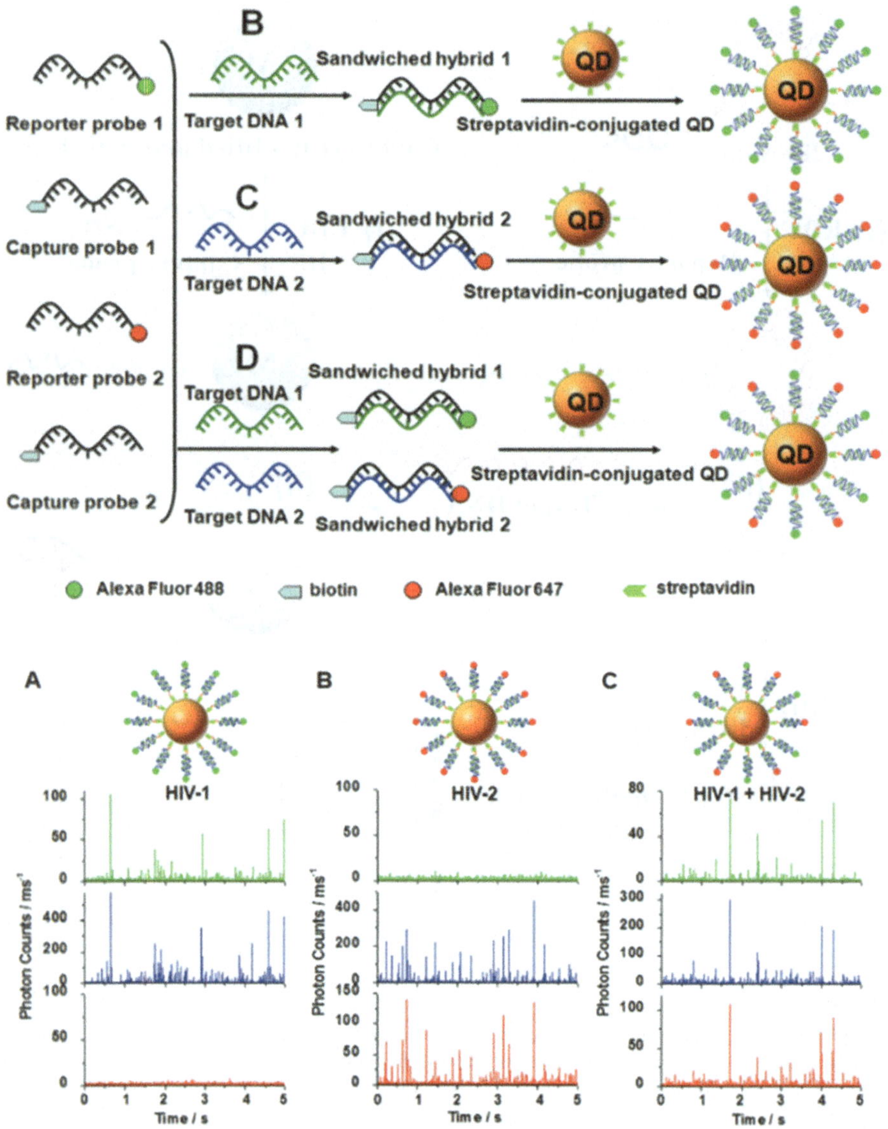

Fig. 5.7 Outline of a single QD-based nanosensor for the detection of multiple DNAs. Representative traces of fluorescence bursts from Alexa Fluor 488, the 605QDs, and Alexa Fluor 647 detected by a single QD-based nanosensor in the presence of HIV-1DNA (**a**), HIV-2 DNA (**b**), and both HIV-1 and HIV-2 DNAs (**c**). Adapted with permission from Han et al. (2001)

HIV-1 assay and Alexa Fluor 647-labeled reporter probe 2 and capture probe 2 for HIV-2 DNA assay. Simultaneous detection of fluorescence eruption from Alexa Fluor 488, Alexa Fluor 647A, and 605QDs (605-nm-emission QDs) may be detected by a single QD-based nanosensor. A single light source of a 488 nm laser was used to excite Alexa Fluor 488 and 605QDs complex (in Alexa Fluor 488-DNA-605QD complex) simultaneously, and the presence of target DNA 1 was denoted as a two-color coincidence signal. As the Alexa Fluor 647-DNA-605QD complex was excited under 488 nm laser, the emission of Alexa Fluor 647 might be observed as a result of FRET from the 605QDs to Alexa Fluor 647 which signify the presence of DNA 2. In this QD-based nanosensor, the QDs played a dual role as a fluorescence pair and a FRET donor for coincidence detection and FRET detection respectively, and also as a local nano concentrator which amplifies the coincidence-related fluorescence signals and the FRET signals. In the absence of target DNA 2, no Alexa Fluor 647 fluorescence signals were detected due to the non-linkage of QDs with Alexa Fluor 647.

5.4.2 Tuberculosis

Tuberculosis (TB) is still an infectious disease and responsible for millions of deaths per year. For its detection, up to 6–8 weeks are required to provide the results in the conventional "culture" methods. An SPR biosensor allows immobilization of nine TB antigens onto the sensor chip based on an array format for the simultaneous detection of antibodies (Dallaire et al. 2015; Prabhakar et al. 2008; Shang-Chen et al. 2012; Silvestri et al. 2015). Conventional photolithography techniques were used to fabricate twenty-five-spot protein arrays. A dropwise addition of 8-mercaptooctanoic acid in ethanol on the surface array for 30 min produced the self-assembled monolayer. Further, a treatment of the surface array was done for 10 min with 400 mM 1-ethyl-3-(3-dimethylaminopropyl) carbodiimide hydrochloride) EDC/100 mM NHS to activate carboxyl groups of 8-mercaptooctanoic acid. For all antigens, a fixed concentration of 50 μg ml^{-1} was sustained and exposed to an array of protein spots for 60 min. The arrays were blocked with 1M ethanolamine for 10 min.

Prabhakar et al. (2008) reported a nucleic acid-based sensor for TB detection based on SPR of gold nanoparticles using Au/DNA and peptide nucleic acid Au/PNA bioelectrodes to detect the sequence specific to *M. tuberculosis*, one-base mismatch, and non-complementary targets. The SPR results of Au/DNA electrode (Fig. 5.8a) showed 203 millidegrees angle change after hybridization with the target (7.3 ng μl^{-1}) due to electrode saturation (Curve i). After the hybridization with one-base mismatched sequence, the change in the angle of about 10.5 millidegrees was observed (Curve ii) due to nonspecific binding. The SPR curves for the PNA/Au electrode are shown in Fig. 5.8b where 251 millidegrees angle change was observed with the sequence (7.3 ng μl^{-1}) due to electrode saturation (Curve i), whereas non-complementary sequence and the one-base mismatch (Curve ii) revealed absence of nonspecific binding (Curve iii) exhibiting negligible binding (error of ~2%). The authors utilized PNA/Au bioelectrode to hybridize with the six-minute sonicated *M. tuberculosis* genomic DNA (Fig. 5.8c). They used the concentration range 5–50 ng μl^{-1} of the genomic DNA with 600 s of association time to observe

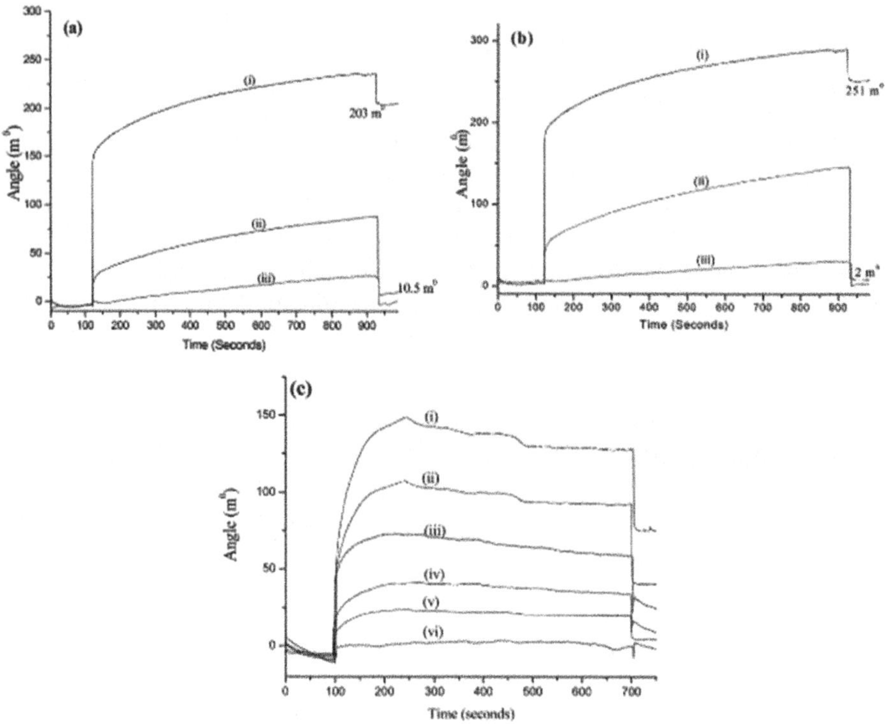

Fig. 5.8 SPR sensor grams of (**a**) Au/DNA bioelectrodes; (**b**) Au/PNA bioelectrodes for hybridization detection with (i) complementary, (ii) one-base mismatch and (iii) non-complementary targets using Au/PNA bioelectrode; and (**c**) PNA/Au bioelectrodes with genomic DNA concentrations of (i) 50.0 ng μl^{-1} (ii) 30.0 ng μl^{-1} (iii) 20.0 ng μl^{-1} (iv) 15.0 ng μl^{-1}, (v) 10.0 ng μl^{-1} and (vi) 5.0 ng μl^{-1}. Adapted with permission from Prabhakar et al. (2008)

the hybridization. The limit of detection for the *M. tuberculosis* genomic DNA was 10.0 ng μl^{-1}.

He et al. (2011) were functionalized the silver nanoparticles for the ultrasensitive detection of *M. tuberculosis* DNA by analyzing the chemiluminescence. Another method is based on a spectroscopic assay that uses silver nanorods for the rapid, sensitive virus detection (Wang et al. 2002). These optical and fluorescent biosensors need transparent sample solutions as color of the sample might interfere with optical and fluorometric signals. So, electrochemical biosensors have gained a lot of attention in the advancement of sequence-specific DNA biosensors (Fan et al. 2010). They are based on the electrical signal produced during the DNA-base pair recognition element.

Zirconium oxide (ZrO_2) nanoparticles deposited on the gold electrode was used to develop a sensitive nucleic acid biosensor for *M. tuberculosis* detection by Das et al. (2010). This biosensor exhibits excellent electrocatalytic properties for the rapid and early identification of *M. tuberculosis* with a limit of detection of 0.065 ng μl^{-1} under 60 s. To covalently immobilize the ss-DNA probe for selective detection of *M. tuberculosis,* a matrix of zirconia nanoparticles immobilized with chitosan has been used. The electrochemical characterization of DNA/CHIT-NanoZrO_2/ITO bioelectrode was done by cyclic voltammetry and differential pulse voltammetry measurements were used to investigate. The sensitivity of the fabricated biosensor was found to be 6.38×10^{-6} A μM^{-1} (Su et al. 2003).

Liu et al. (2014) reported a DNA biosensor based on electrochemical detection with excellent sensitivity. A reduced graphene oxide-gold nanoparticles electrode has been utilized as a sensing platform. A probe-labeled gold-Polyaniline (Au–PANI) nanocomposite was used to amplify the signals. The sensitivity towards *M. tuberculosis* was found to be in the range of 1.0×10^{-5} to 1.0×10^{-9} M. Tsai (2017) developed an analytical paper-based device using gold nanoparticles which works on colorimetric sensing principle without any sophisticated instrument. Further, the variation in color of gold nanoparticles colloid was observed when single-stranded DNA probe molecules hybridized with targeted double-stranded TB DNA. For TB DNA sequences, the detection limit was found to be 1.95×10^{-2} ng ml^{-1}, with a linear dynamic range of 1.95×10^{-2} to 1.95×10^{1} ng ml^{-1}. It should be noted that the turnaround time for this paper-based analytical sensing device can be realistically concluded in 60 min after DNA extraction.

Phunpae et al. (2014) reported a sandwich enzyme-linked immunosorbent assay (ELISA) platform where antibodies to Antigen 85 of *M. tuberculosis* were created based on the colorimetric technique for *M. tuberculosis* detection having a limit of detection 8 ng ml^{-1} in the linear array of 8–400 ng ml^{-1}. They achieved 89.6% sensitivity and 94% specificity by detecting the Ag85 proteins in culture filtrate with a faster diagnostic result than the standard mycobacterial culture method. Some of the tested samples of *M. tuberculosis* gave negative results with ELISA while they were positive samples. The researchers assumed the low amount of Ag85 in collected culture filtrates, is responsible for the wrong results by the developed ELISA at day 28. This system can detect 25%, 50%, 80%, and 90% of TB positive samples by days 3, 7, 14, and 28, respectively, compared to the MGIT 960 system or the standard mycobacterial method (Palomino 2005). In another report, a new fluorimetry based sandwich ELISA technique was developed for evaluating Ag85B antigen using Au nanorods and QDs. This latest technique has limit of detection 0.013 ng ml^{-1} of Ag85B in the linear range of 0.013 ng ml^{-1} to 1.0 ng ml^{-1} (Tsai 2017). Das et al. (2011) reported a multiwalled carbon nanotube (MWCNT) based biosensor for *M. tuberculosis* detection. In this study, an Indium-tin-oxide (ITO) coated glass substrate was coated with MWCNT and Zirconia through a controllable electrophoretic deposition to form Nano-ZrO_2-CNT films. The Nano-ZrO_2-CNT/ITO electrode was used for detecting ss-DNA specific to *M. tuberculosis*. The ss-DNA-NanoZrO_2-CNT/ITO bioelectrode has been found to detect target DNA selectively for the concentrations ranging from 1×10^{-2} to 1×10^{-8} mM with an improved limit of detection of 0.01 nM.

5.4.3 Hepatitis

Zhu et al. (2015) have reported a FRET biosensor for detection hepatitis B virus (HBV) DNA. The biosensor was fabricated with AuNPs as an acceptor and poly (ethylenimine) modified upconversion nanoparticles (NH_2-UCNPs) as an energy donor. Two strands of ss-DNA were conjugated with AuNPs and NH_2-UCNPs. The hybridization between UCNPs and AuNPs-conjugated complementary DNA leads to upconversion luminescence quenching due to the FRET process. The addition of target DNA released AuNPs from the UCNPs surface and the upconversion luminescence was restored due to the binding of stable ds-DNA on UCNPs. The designed biosensor demonstrated great selectivity and sensitivity with the potential for use in clinical analysis of HBV and others.

Riedel et al. (2016) reported a method for the detection of protein biomarkers using a plasmonic biosensor. Clinical serum specimens were analyzed using bio-receptor functionalized poly[(*N*-2-hydroxypropyl) methacrylamide)co (carboxybetaine methacrylamide)] brushes (Fig. 5.9). It was the first time that this method allowed direct detection of anti-HBs against HBV surface antigen in clinical serum specimens utilizing SPR and also exceptional resistance to fouling was provided. The SPR biosensor can detect positive and negative anti-HBs clinical specimens within 10 min. The required amount of sample for a single analysis was only 15 μL. The diagnosis obtained by SPR was compared with the results obtained from ELISA, which was carried out by an independent certified laboratory on the same samples. A glycine buffer can regenerate the biosensor with proper treatment.

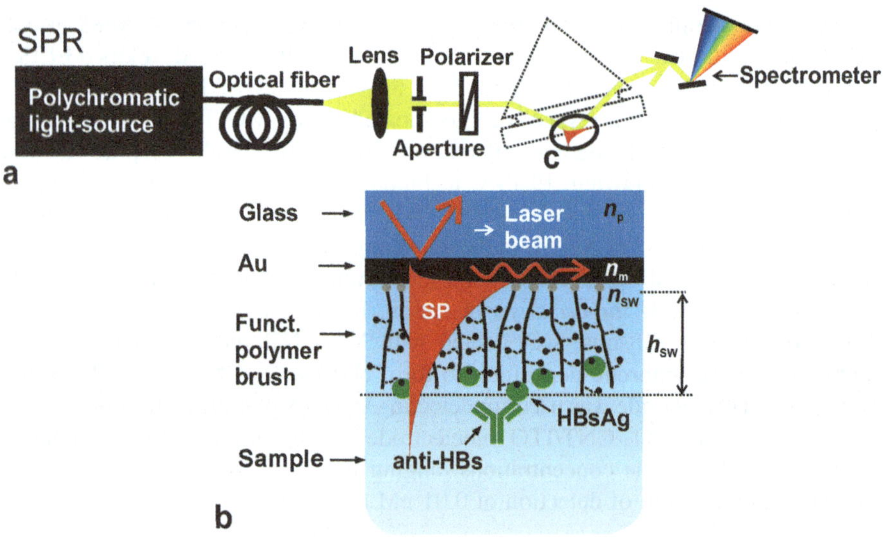

Fig. 5.9 The surface plasmon resonance setup (**a**). Sensor chip with poly(HPMA-co-CBMAA) brush functioning as a binding matrix for direct detection of the anti-HBs target analyte (**b**). Adapted with permission from Riedel et al. (2016)

The authors reported the possibility to regenerate the biosensor after 3 sets of the measurement with up to 95% efficiency.

In a recent study, a modified graphite electrode with poly(4-aminophenol) with a special oligonucleotide probe was fabricated to detect HBV specific DNA sequence. Differential pulse voltammetry was employed to assess the probe containing a modified electrode with or without HBV specific DNA sequence's incubation. The direct monitoring of oxidizable DNA bases or indirect detection of ethidium bromide as an indicator for the hybridization procedure was used to complete the detection. The direct monitoring of oxidizable DNA biosensor had the limit of detection of 2.61 nmol l^{-1}. The indirect detection was a worthwhile feature for detecting point mutation-related diseases through differentiating mismatches in the case of ethidium bromide. The increase in current with the concentration of the complementary target indicated the hybridization with the complementary target and on the surface accumulation of ethidium bromide. A central phenanthridine ring system is in ethidium bromide intercalated into double-stranded DNA. A linear relationship of current versus concentration of the complementary target was observed between 0.18×10^{-6} and 1.8×10^{-6} mol l^{-1} (Castro et al. 2014).

Mandli et al. (2017) developed an immunosensor for the detection of Hepatitis A virus (HAV) antigen through immobilization of HAV on the carbon nano-powder paste electrode surface (Fig. 5.10). In order to monitor the target HAV antigen, a secondary antibody labeled with peroxidase was utilized. An amperometric current signal was detected on the addition of hydroquinone and hydrogen peroxide as a redox mediator. A plot of amperometric signal versus HAV antigen concentration resulted in the calibration plot with a limit of detection for HAV antigen 26×10^{-5} IU ml^{-1} with a linear range of 2×10^{-4} IU ml^{-1} to 5×10^{-3} IU ml^{-1}. The detection of HAV antigen in aqueous samples demonstrates high selectivity and stability of the immunosensor towards the HAV antigen (Mandli et al. 2017).

The detection of HBV antigen was performed through self-assembled AuNPs with solgel network by using a potentiometric immunosensor. The amplification of the signal as well as the sensitivity of the immunosensor can be achieved by the hybridization of thiol-containing solgel network and AuNPs. The AuNPs were absorbed onto the thiol group by immersing the gold electrode into hydrolyzed (3-mercaptopropyl) trimethoxysilane solgel to assemble a 3D-silica gel. In turn, the surface antibodies of HBV were assembled onto the gold nanoparticles surface. The electrochemical impedance spectroscopy and cyclic voltammetry were used to evaluate the self-assembling procedure. The authors claimed a linear range of 4–960 ng ml^{-1} with long-time stability, great reproducibility and extreme sensitivity and a limit of detection 1.9 ng ml^{-1} (Tang et al. 2006).

Narang et al. (2016) developed a zeolite nanocrystal and MWCNTs based genosensor for the polymerase chain amplified HBV DNA detection in blood serum samples. The hybridization between target and probe ss-DNA resulted in decreased current through the interaction of methylene blue with 3′G (free guanine) of ss-DNA. The peak currents increased with the increasing scan rate and the plot tends to level off at higher scan rates 100–500 mV s^{-1}. In Fig. 5.11a, b CV of nanostructure modified electrode was recorded at different scan rates from

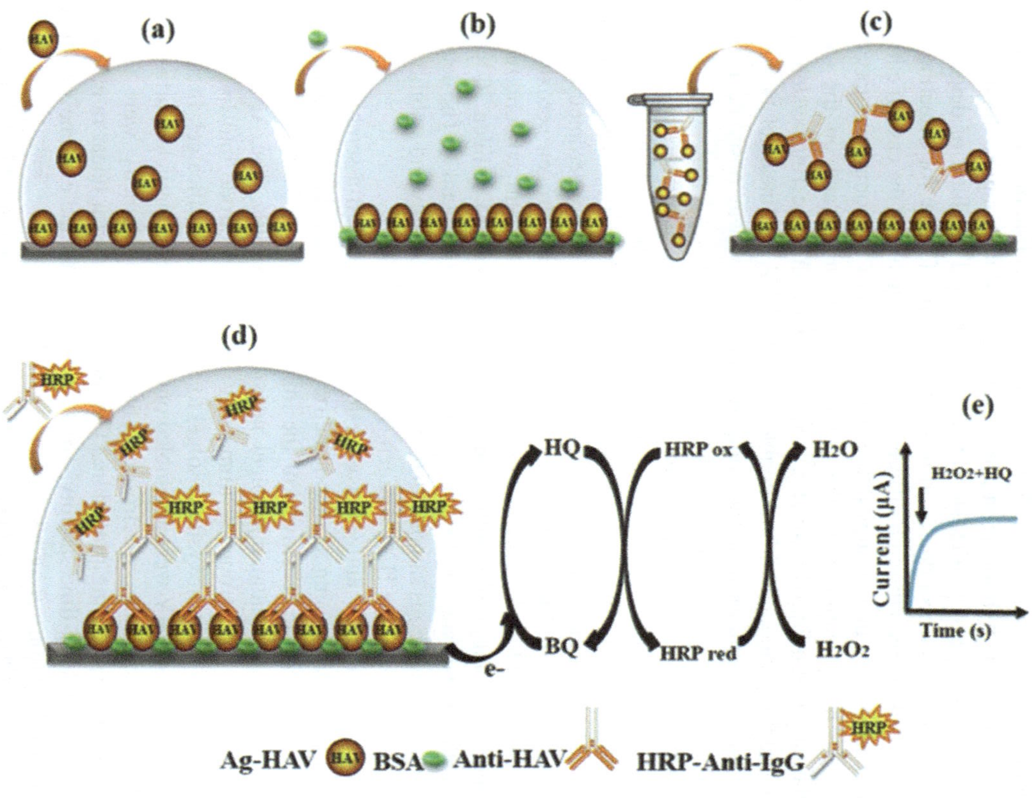

Fig. 5.10 The protocol involved for the preparation of the electrochemical HAV competitive antibody immunosensor. Adapted with permission from Mandli et al. (2017)

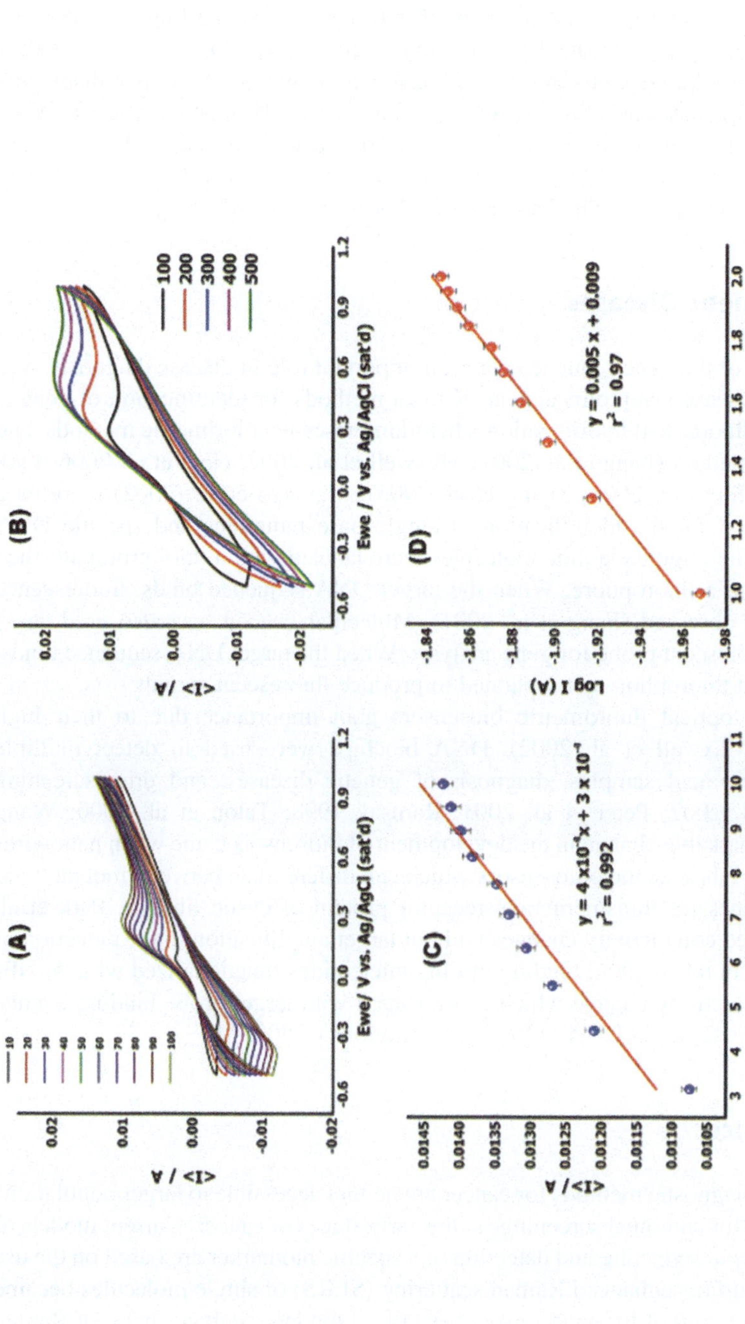

Fig. 5.11 **(a)** CV obtained at ss-DNA/MWCNT-zeolite/FTO for scan rates 10–100 mV s^{-1} in 0.50 M acetate buffer (pH 4.80) with 20 mM NaCl **(b)** CV obtained at ss-DNA/MWCNT-zeolite/FTO for scan rates 100–500 mV s^{-1} **(c)** the dependency of peak currents on the square root of potential sweep rates in a wide range of 10–100 mV s^{-1} **(d)** dependence of log of peak current on log v (mV s^{-1}). Adapted with permission from Narang et al. (2016)

10 to100 mV s^{-1} as well as from 100 to 500 as mV s^{-1}. No shifting of peaks with increasing scan rates confirmed the stability of the sensor (Fig. 5.11c). The slow electron transfer kinetics on the CV time scale was shown by the sensor displaying peak current increases as a function of the square root of the scan rate, qualitatively. In Fig. 5.11d the plot of log I versus log v confirms the diffusion-controlled behavior. The limit of detection of genosensor was found to be 50 copies ml^{-1} and 99% accuracy was achieved in the detection of HBV in the patient's blood.

5.4.4 Genetic Diseases

The analysis of the gene sequence plays an important role in disease diagnosis even before any disease symptoms appear. Known methods for identification of genetic diseases are fluorescent hybridization, chemiluminescent, colorimetric methods, and label-based method (Fang et al. 2001; Maxwell et al. 2002; Negi et al. 2006; Park et al. 2002; Su et al. 2003; Zhang et al. 2005). Maxwell et al. (2002) reported a unique method for the identification of single base mutations and specific DNA sequences. The oligonucleotide molecules were labeled with a thiol group attached to AuNPs and a fluorophore. When the target DNA sequence binds, fluorescence signals were observed (Fang et al. 2001). Molecular beacon was also used as an excellent fluorescent probe for gene analysis. When the target DNA sequence binds, quencher and fluorophore get separated to produce fluorescent signals.

Later on, optical fluorometric biosensors gain importance due to their high sensitivity (Maxwell et al. 2002). DNA biochips were used to detect multiple genes, nucleic acid samples, diagnosis of genetic diseases, and drug screening (Marks et al. 2007; Peter et al. 2001; Ramsay 1998; Taton et al. 2000; Wang 2000). A remarkable change in the development of biosensors came when nanowires were used to fabricate nanobiosensors which can differentiate between mutant, wild type, and genes for transmembrane receptor protein of cystic fibrosis. Park et al. (2002) utilized conductivity changes without target amplification in the detection of target DNA. In this method binding of oligonucleotides functionalized with AuNPs leads to conductivity change which is associated with target-probe binding events. This method can detect DNA at low concentrations of 500 fM.

5.4.5 Cancer

Traditional diagnostic methods for cancer are neither accessible to larger populations nor practical for continual screenings at the early stages of cancer. Current models of high-throughput screening and detection of a specific biomarker are based on the use of ELISA. Surface-enhanced Raman scattering (SERS) of single molecules became a major detection tool for nanosensors because of the large enhancement of Raman scattering on the order of 10^{14}–10^{15} by metal or core-shell nanoparticles (Nie and Emory 1997). This Raman scattering enhancement factor is attributed to the localized surface plasmon resonance modes (LSPR) at the surface of the nanoparticle, which increases the density of states at Stokes-shifted wavelengths by focusing the energy to the nanoparticle. Nie and colleagues have advanced the SERS field in nanotechnology (Anker et al. 2008) as well for in vivo cancer detection (Qian and Nie 2008).

A no-contact method was proposed by utilizing a Raman spectrometer outside the live animal for in vivo sensing through SERS tags that target cancer cells (Qian et al. 2008). Certain chromophores can adsorb to PEGylated AuNPs by electrostatic interactions and maintain Raman enhancement factors. Target groups and PEG chains are bound to the AuNPs surface via thiols and hetero-functional PEG. Qian et al. (2008) encapsulated AuNPs with PEG, antibodies, and small-molecule Raman reporters to non-invasively detect high SERS signals targeted to cancer cells in vivo. The specific biomolecular targeting and detection using SERS nanotags by AuNPs or AgNPs labeled with fluorescent dyes or surfactants was achieved (Huang et al. 2007; Qian et al. 2008; Yu et al. 2007). The limit of detection with SERS detection has reached 10^{-21} M for DNA and RNA (Cao et al. 2002) and fM level for prostate-specific antigen (Grubisha et al. 2003).

Stern et al. (2010) have detected two cancer antigens from a 10 μL whole blood sample in under 20 min by using a microfluidic purification chip to concentrate the target before electrical detection (Fig. 5.12). The reasons for this remarkable detection are the FET device and the purification chip that concentrates the sample. The microfluidic chip captures various types of biomarkers from blood, washes them, and releases the markers into a purified buffer for electronic sensing. UV irradiation on the photo-cleavable cross-linker between the device and biomarkers of interest was used to release the markers. The prevention of high salt concentration through pre-processing allows cheaper detectors with clinically relevant samples by minimizing the interfering with the electronic signals and hence.

The magnetic nanosensors are simple and effective to use and can be useful to outline cancer cells. A magnetic nanosensor bearing carbohydrates was developed to utilize the MRI technique to describe the carbohydrate-binding characteristics of cancer cells quantitatively and qualitatively which can simplify both therapeutic tools and molecular diagnostics for cancer (Stern et al. 2010). The chip-based micronuclear magnetic resonance system is a powerful tool that offers unique advantages in molecular profiling of the cancer cell (Haun et al. 2011; Lee et al. 2008). The system consists of micro coils for radio-frequency excitation and nuclear magnetic resonance signal detection, a portable magnet, an on-board nuclear magnetic resonance spectrometer, and microfluidic networks. Super paramagnetic iron oxide nanoparticles were conjugated with antibodies to each target, followed by incubation with cancer cells. The significant differences in T_2 relaxation time can be observed for various cancer cells using this system. The proposed approach with high sensitivity, specificity, and high-throughput shows the possibility for early cancer diagnosis in the clinic (Haun et al. 2011).

DNA hybridization methods have attracted attention in the management of cancer. Various techniques have been used for DNA hybridization such as electrochemical (Wang et al. 2003), fluorescence (Zhao et al. 2003), surface plasmon resonance spectroscopy (Peterlinz et al. 1997), enzymatic (De et al. 1996), and

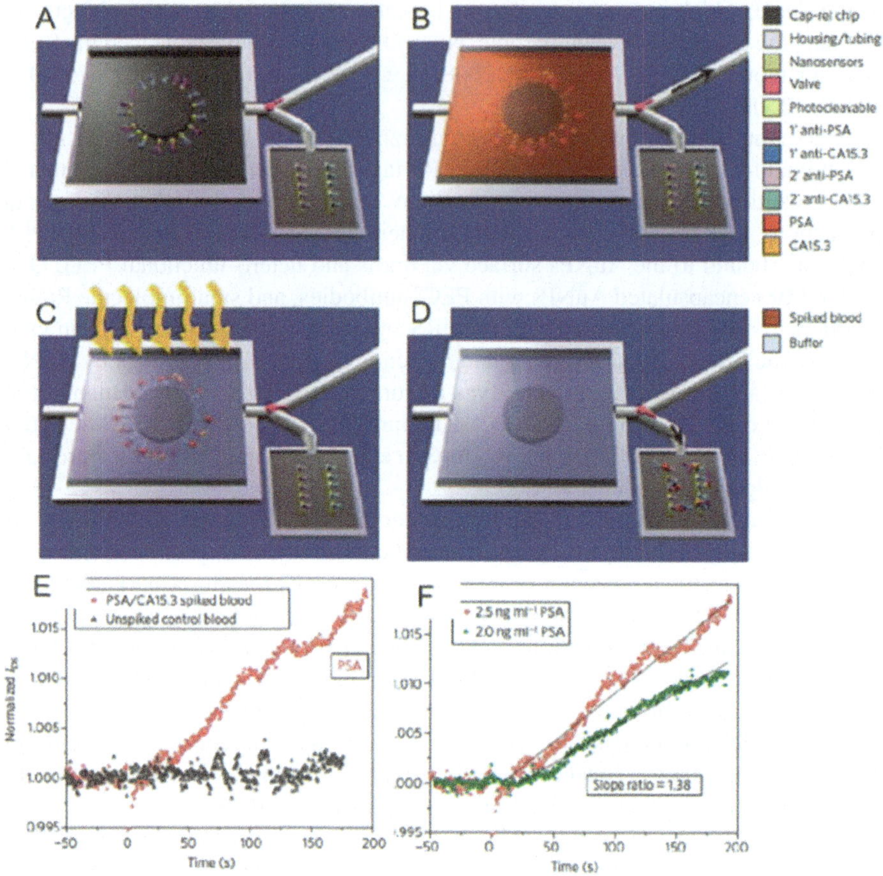

Fig. 5.12 A pre-processing electrical detection method. (**a**) Primary antibodies are bound to the sensor via a photo-cleavable cross-linker. (**b**) Whole blood is injected into the chip (black arrow) and biomarkers bind to the device. (**c**) The probe is washed and then UV irradiation (orange waves) is applied to cleave the linker between the captured biomarker and sensor. (**d**) Finally, the antibody–antigen complexes are washed out of the sensor for detection. (**e**) Response to an anti-prostate-specific antigen (PSA) purified from a blood sample, initially containing 2.5 ng ml^{-1} PSA compared with no protein. (**f**) Normalized response to different concentrations of PSA. Adapted with permission from Stern et al. (2010)

colorimetric (Liu and Lu 2004). Electrochemical transducers are more preferable due to simplicity, low cost, high sensitivity, and portable (Cai et al. 2003; Wang 1993). For the immobilization of single-stranded probes there were different methods that were used such as self-assembly, entrapment, covalent binding, and biotin-avidin interactions. Zhu et al. (2004) reported that the electrodeposited ZnO_2 thin films on gold electrodes can be used in the fabrication of nanobiosensors based on DNA immobilization. Similarly, Feng et al. (2006) reported a composite film as a matrix for the immobilization of DNA sequence for colorectal cancer. The proposed electrochemical sensor has a limit of detection from 1.59×10^{-11} M to 1.16×10^{-7} M and the ability to differentiate a complementary target sequence and base mismatch sequence.

5.4.6 Infectious Diseases

Infectious diseases are the most critical problems of the current era and have significant challenges. The evolution of virulent new infectious agents, such as Zika and Chikungunya, or H1N1 has been accelerated in recent years. The evolution of resistant strains of known bacteria is a big challenge because they are passive to present antibiotics. The resistance in bacteria is generally acquired by repeated use of antibiotics. Currently, the world needs a rapid approach to identify the infections at the early stages and to provide treatment that is harmful to the bacteria but not to humans. The early vaccination is an effective approach to prevent the spread of the infection and can vanish certain diseases globally. But the identification of effective antigen and type of activation of the immune system to promote long-term protection is still a challenge.

Chen et al. (2015) used the optical properties and narrow wavelength fluorescence of QDs to create a Barcoding approach toward disease detection. The QDs were encapsulated into larger polystyrene microspheres to make QDs highly efficient, bright, and stable fluorophores. The encapsulated QDs were modified with a capture DNA strand that binds to a known DNA sequence associated with a given pathogen; a different QD color is associated with each DNA type that can be detected by measuring the microbead fluorescence (Adekoya et al. 2006).

Furthermore, Driskell et al. (2011) used AuNPs to detect the influenza virus <100 TCID 50 ml^{-1} within 30 min. Le et al. (2014) developed a platform utilizing AuNPs for the visual detection of viral particles as little as 3×10^8 in number. To achieve the specific affinity with strains of human influenza, AuNPs were assembled with RNA aptamers. Aptamers are cheaper than the targeting mAbs, as they are 1/1000th the price per molecule. The functionalized nanoparticles form a gold nanoshell around the virus in the presence of viral contaminants. The viral contaminants get settled down due to sedimentation and can be facilitated with the centrifuge. A visual sedimentation of the nanoparticle-coated virus, after centrifugation at 2000 g, was reported in 1 ml of human respiratory specimens.

Ansari et al. (2009) reported label-free nanobiosensor for *Neisseria gonorrhoeae* based on nano zinc oxide film, dip-coated on indium-tin-oxide (ITO). The immobilization of thiolated single-stranded oligonucleotide (ss-DNA) probe on Au electrode leads to novel electrochemical sensors for *Neisseria meningitides*. The electrode was stable for nearly 120 days at 4 °C. The sensitivity of ds-DNA)/Au electrode was found to be 115.8 μA ng^{-1}. Later on, a different approach was proposed by Singh et al. (2010b) for the detection of *N. gonorrhoeae* by fabricating nanobiosensors. A nanobiosensor has been fabricated based on a multi-copy gene of *N. gonorrhoeae* functionalized nanostructured–polyaniline coated onto ITO-coated-glass plate. Avidin-biotin, a cross-linking biopolymer, was used in this biosensor.

The limit of detection of this electrode was up to 0.5×10^{-5} M of the complementary target within 60 s of hybridization at 25 °C by differential pulsed voltammetry. The electroactive DNA hybridization indicator was methylene blue. The fabricated sensor can distinguish *N. gonorrhoeae* from *N. meningitides* and *E. coli* in cultures from urethral swabs of patients (Hlavata et al. 2014).

Fe_3O_4/chitosan deposited on ITO was used by Singh et al. (2010b) for the detection of gonorrhoeae. These nanocomposite films were stable and can specifically detect DNA. This method provides with low detection limit with a high-throughput rate. Chitosan doped MWCNTs were also used to develop the biosensor for gonorrhoeae detection. The electroactive area was enhanced by CNTs up to threefolds. The detection limit was found to be 1×10^{-16} M with a linear range of 1×10^{-6} M to 1×10^{-17} M. The response time was found to be 60 s and the sensor was stable for about 4 months.

Singh et al. (2010a) have used PANI/CNTs hybrid nanocomposite for selective detection of gonorrhoeae by immobilizing 5′-amino-labeled *N. gonorrhoeae* probe and glutaraldehyde was used as a cross-linker. The biosensor exhibits a limit of detection 2×10^{-17} M and linear range from 1×10^{-6} M to 1×10^{-7} M. This biosensor showed high sensitivity, good reproducibility, rapid response, and also long-term stability.

Tam et al. (2009) proposed label-free and direct detection of influenza virus (type A) by covalently immobilizing probe DNA on MWCNTs. Raman and FTIR spectra were used to observe the covalent bonding between amines and phosphate groups of the DNA. The limit of detection was found to be 0.5 nM. Semiconductor nanoparticles such as CdSe, CdS, PbS, and ZnS have attracted a lot of attention for electrochemical biosensing. They are widely used due to their unique size-tuneable fluorescent properties (He et al. 2011). The semiconductor nanoparticle tags were first used by Wang's group for the assay of electrochemical DNA hybridization (Wang et al. 2002). Fan et al. (2010) fabricated an electrochemical DNA biosensor based on semiconductor quantum dots for the detection of the avian influenza virus. DNA/CdSe modified GCE was used to develop label-free or indicator-free DNA hybridization detection using electrochemical techniques.

Jagriti et al. (2018) reported a genosensor for the selective detection of *E. coli* Shigella toxin DNA. The linear detection range was found to be 1 fM–100 μM with 1 fM limit of detection. This genosensor serves as a promising platform for the detection of *E. coli* toxin.

Patel et al. (2015) reported a photoluminescence based genosensor for cholera detection using magnesium oxide nanoparticles. A linear range of response in the developed genosensor was observed from 100 to 500 ng μl^{-1} with a sensitivity of 1.306 min g^{-1} and a limit of detection 3.133 ng μl^{-1}. Chitosan modified magnesium oxide nanoparticles were also explored for the detection of cholera (Patel et al. 2013b).

Patel et al. (2013a) also reported the magnesium oxide grafted CNTs based impedimetric genosensor for biomedical applications. The electrochemical impedance spectroscopy (EIS) results reveal the sensitivity of 3.87 Ω ng^{-1} cm^{-2}, a detection limit of ~21.70 ng μl^{-1} in the linear range of 100–500 ng μl^{-1}. In another study, they reported nanostructured zirconium oxide nanoparticles for bacterial detection (Solanki et al. 2011).

5.5 Conclusions/Outlook

The nano patterning of materials can be done by developing DNA-based nanobiosensors using new methods. The most important clinical applications of nanosensors are in the areas of detection of genetic, immunodeficiency, infectious diseases, cancer diagnosis, etc. To date, optical nanosensors have been used to detect a broad variety of analytes, ranging from small chemicals, proteins, nucleic acids, bacterial, and viral pathogens. The role of PCR can be eliminated by using DNA biostrip and biochip technologies. New reliable devices are needed and the improvement of the existing devices is required for future biosensors in order to allow amplification, superior transduction, processing, and conversion of the biological signals. The huge role of nanotechnology cannot be denied in the development of biosensors for toxin detection. Nanomaterials offer new prospects to enhance the performance of the electrochemical biosensors for sensing. In spite of recent advancements in the development of low-fouling functional coatings, the nonspecific adsorption of non-target molecules from a sample matrix remains the main challenge that needs to be overcome to enable plasmonic biosensors to enter routine analytical practice. Considering all of these aspects, it can be stated that nano-biosensors offer the possibility of diagnostic tools with increased sensitivity, specificity, and reliability for in vivo and in vitro analytical applications. However, nano-biosensors still need to achieve the confidence of potential users, especially considering that the commercialization of new devices is the aim of nanobiosensor technology development.

References

Abdul-Quader AS, Feelemyer J, Modi S, Stein ES, Briceno A, Semaan S, Horvath T, Kennedy GE, Des Jarlais DC (2013) Effectiveness of structural-level needle/syringe programs to reduce HCV and HIV infection among people who inject drugs: a systematic review. AIDS Behav 7:2878–2892

Abu-Salah KM, Zourob MM, Mouffouk F, Alrokayan SA, Alaamery MA, Ansari AA (2015) DNA-based nanobiosensors as an emerging platform for detection of disease. Sensors 15:14539–14568

Andersen FF, Stougaard M, Jorgensen HL, Bendsen S, Juul S, Hald K, Andersen AH, Koch J, Knudsen BR (2009) Multiplexed detection of site specific recombinase and DNA topoisomerase activities at the single molecule level. ACS Nano 3:4043–4045

Anker JN, Hall WP, Lyandres O, Shah NC, Zhao J, Van Duyne RP (2008) Biosensing with plasmonic nanosensors. Nat Mater 7:442–453

Ansari AA, Singh R, Sumana G, Malhotra BD (2009) Sol-gel derived nano-structured zinc-oxide film for sexually transmitted disease sensor. Analyst 134:997–1002

Ansari MIH, Hassan S, Qurash A, Khanday FA (2016) Microfluidic-integrated DNA nanobiosensors. Biosens Bioelectron 85:247–260

Balaji P, Brittany B, Palniappan S, Madhusudhana R, Joseph I (2011) Gold nanoprobes for theranostics. Nanomedicine 6:1787–1811

Bhushan B (2010) MEMS/NEMS and BioMEMS/BioNEMS: materials devices and biomimetics. In: Bhushan B (ed) Springer handbook of nanotechnology. Springer, New York, pp 1663–1740

Cai H, Cao X, Jiang Y, He PG, Fang YZ (2003) Carbon nanotube-enhanced electrochemical DNA biosensor for DNA hybridization detection. Anal Bioanal Chem 375:287–293

Caires AJ, Mansur HS, Mansur AAP, Carvalho SM, Lobato ZIP, dos Reis JKP (2019) Gold nanoparticle-carboxymethyl cellulose nano colloids for detection of human immunodeficiency virus type-1 (HIV-1) using laser light scattering immunoassay. Colloids Surf B Biointerfaces 177:377–388

Cao YC, Jin R, Mirkin CA (2002) Nanoparticles with Raman spectroscopic fingerprints for DNA and RNA detection. Science 297:1536–1540

Castro ACH, França EG, de Paula LF, Soares MM, Goulart LR, Madurro JM, Brito-Madurro AG (2014) Preparation of genosensor for detection of specific DNA sequence of the hepatitis B virus. Appl Surf Sci 314:273–279

Chen PY, Chung MT, McHugh W, Nidetz R, Li YW, Fu JP, Cornell TT, Shanley TP, Kurabayashi K (2015) Multiplex serum cytokine immunoassay using nanoplasmonic biosensor microarrays. ACS Nano 9:4173–4181

Cheng X, Irimia D, Dixon M, Ziperstein JC, Demirci U, Zamir L, Tompkins RG, Toner M, Rodriguez WR (2007) A microchip approach for practical label-free CD4+ T-cell counting of HIV-infected subjects in resource-poor settings. J Acquir Immune Defic Syndr 45:257–261

Dahman Y, Radwan A, Nesic B, Isbister J (2017) Nanotechnology and functional materials for engineers. Nanosensors 2017:67–91

Dallaire AM, Patskovsky S, Vall A, Meunier M (2015) Electrochemical plasmonic sensing system for highly selective multiplexed detection of biomolecules based on redox nanoswitches. Biosens Bioelectron 71:75–81

Darestani-Farahania M, Faridboda F, Ganjal MR (2018) A sensitive fluorometric DNA nanobiosensor based on a new fluorophore for tumor suppressor gene detection. Talanta 190:140–146

Das M, Sumana G, Nagarajan R, Malhotra BD (2010) Zirconia based nucleic acid sensor for Mycobacterium tuberculosis detection. Appl Phys Lett 96:13

Das M, Dhand C, Sumana G, Srivastava AK, Vijayan N, Nagarajan N, Malhotra BD (2011) Zirconia grafted carbon nanotubes based biosensor for M. tuberculosis detection. Appl Phys Lett 99:14

De L, Campbell T, Heller C (1996) Direct enzyme-amplified electrical recognition of A30-base model oligonucleotide. J Am Chem Soc 118:5504–5505

Doucey M, Carrara S (2019) Nanowire sensors in cancer. Trends Biotechnol 37:86–99

Driskell JD, Jones CA, Tompkins SM, Tripp RA (2011) One-step assay for detecting influenza virus using dynamic light scattering and gold nanoparticles. Analyst 136:3083

Erickson D, Mandal S, Yang AHJ, Cordovez B (2008) Nanobiosensors: optofluidic, electrical and mechanical approaches to biomolecular detection at the nanoscale. Microfluid Nanofluid 4:33–52

Fan H, Ju P, Ai S (2010) Controllable synthesis of CdSe nanostructures with tunable morphology and their application in DNA biosensor of Avian Influenza Virus. Sensors Actuators B Chem 149:98–104

Fang X, Li JJ, Perlette J, Tan W, Wang K (2001) Molecular bacons: novel fluorescent probes. Anal Chem 72:747A–753A

Feng KJ, Yang YH, Wang ZJ, Jiang JH, Shen GL, Yu RQ (2006) A nano-porous CeO2/chitosan composite film as the immobilization matrix for colorectal cancer DNA sequence-selective electrochemical biosensor. Talanta 70:561–565

Gleiter H (1995) Nanostructured materials: state of the art and perspectives. Nanostruct Mater 6:3–14

Global AIDS Update (2016) Enormous gains, persistent challenges. www.who.int/hiv/pub/arv/global-AIDS-update-2016_en.pdf

Grubisha DS, Lipert RJ, Park HY, Driskell J, Porter MD (2003) Femtomolar detection of prostate-specific antigen: an immunoassay based on surface-enhanced Raman scattering and immunogold labels. Anal Chem 75:5936–5943

Han MY, Gao XH, Su JZ, Nie SM (2001) Quantum-dot-tagged microbeads for multiplexed optical coding of biomolecules. Nat Biotechnol 19:631–635

Haun JB, Castro CM, Wang R, Peterson VM, Marinelli BS, Lee H, Weissleder R (2011) Micro-NMR for rapid molecular analysis of human tumor samples. Sci Transl Med 3:71ra16

He B, Morrow TJ, Keating CD (2008) Nanowire sensors for multiplexed detection of biomolecules. Curr Opin Chem Biol 12:522–528

He Y, Liu D, He X, Cui H (2011) One-pot synthesis of luminol functionalized silver nanoparticles with chemiluminescence activity for ultrasensitive DNA sensing. Chem Commun 47:10692–10694

Hlavata L, Vyskocil V, Benikova K, Borbelyova M, Labuda J (2014) DNA-based biosensors with external Nafion and chitosan membranes for the evaluation of the antioxidant activity of beer, coffee and tea. Cent Eur J Chem 12:604–611

Huang X, El-Sayed IH, Qian W, El-Sayed MA (2007) Cancer cells assemble and align gold nanorods conjugated to antibodies to produce highly enhanced, sharp, and polarized surface Raman spectra: A potential cancer diagnostic marker. Nano Lett 7:1591–1597

Hui-Fen W, Kavita A, Kamlesh S, Yi-Hsien L (2010) On particle ionization/enrichment of multifunctional nanoprobes: washing/separation-free, acceleration and enrichment of microwave-assisted tryptic digestion of proteins via bare TiO2 nanoparticles in ESI-MS and comparing to MALDI-MS. J Mass Spectrom 45:1402–1408

Jagriti N, Annu M, Roberto P, Alekhya VV, Shikha W, Chandra SP, Manika K (2018) Development of MoSe2 nano-urchins as a sensing platform for a selective bio-capturing of Escherichia. coli Shiga Toxin DNA. Biosensors 8:77

Jokerst JV, Floriano PN, Christodoulides N, Simmons GW, McDevitt JT (2008) Integration of semiconductor quantum dots into nano-bio-chip systems for enumeration of CD4+ T cell counts at the point-of-need. Lab Chip 8:2079–2090

Khan AAP, Khan A, Rahman MM, Asiri AM, Oves M (2016) Lead sensors development and antimicrobial activities based on graphene oxide/carbon nanotube/poly(O-toluidine) nanocomposite. Int J Biol Macromol 89:198–205

Le TT, Adamiak B, Benton DJ, Johnson CJ, Sharma S, Fenton R, McCauley JW, Iqbal M, Cass AEG (2014) Aptamer-based biosensors for the rapid visual detection of flu viruses. Chem Commun 50:15533–15536

Lee H, Sun E, Ham D, Weissleder R (2008) Chip-NMR biosensor for detection and molecular analysis of cells. Nat Med 14:869–874

Liao L, Wang S, Xiao J, Bian X, Zhang Y, Scanlon MD, Hu X, Tang Y, Liu B, Girault HH (2014) A nanoporous molybdenum carbide nanowire as an electrocatalyst for hydrogen evolution reaction. Energy Environ Sci 7:387–392

Liu J, Lu Y (2004) Calorimetric biosensor based on DNAzyme-assembled gold nanoparticles. J Fluoresc 14:343–354

Liu C, Xiang G, Liu L, Liu F, Pu X (2014) An electrochemical DNA biosensor for the detection of Mycobacterium tuberculosis, based on signal amplification of graphene and gold nanoparticle–polyaniline nanocomposites. Analyst 139:5460–5465

Malik P, Katyal V, Malik V, Asatkar A, Inwati G, Mukherjee TK (2013) Nanomaterials nanobiosensors: concepts and variations. Int Sch Res 2013:1–9

Mandli J, Attar A, Ennaji MM, Amine A (2017) Indirect competitive electrochemical immunosensor for hepatitis a virus antigen detection. J Electroanal Chem 799:213–221

Marks RS, Lowe CR, Cullen DC, Weetal HH, Karube I (2007) Handbook of biosensors biochips. Wiley, Chichester

Mathers BM, Degenhardt L, Ali H, Wiessing L, Hickman M, Mattick RP, Myers B, Ambekar A, Strathdee SA (2009) HIV prevention, treatment, and care services for people who inject drugs: a systematic review of global, regional, and national coverage. Lancet 375:1014–1028

Maxwell DJ, Taylor JR, Nie S (2002) Self-assembled nanoparticle probes for recognition and detection of biomolecules. J Am Chem Soc 124:9606–9612

Narang J, Singhal C, Malhotra N, Narang S, Gupta R, Kansal R, Pundir C (2016) Impedimetric genosensor for ultratrace detection of hepatitis B virus DNA in patient samples assisted by zeolites and MWCNT nano-composites. Biosens Bioelectron 86:566–574

Negi SS, Anand R, Pasha ST, Basir SF, Gupta S, Khare S, Lal S (2006) Detection of M. tuberculosis in clinical samples of diversified nature by IS6110 based PCR. J Commun Dis 38:325–332

Nie S, Emory SR (1997) Probing single molecules and single nanoparticles by surface-enhanced Raman scattering. Science 275:1102–1106

Ovadekova R, Jantova S, Letasiova S, Stepanek I, Labuda J (2006) Nanostructured electrochemical DNA biosensors for detection of the effect of berberine on DNA from cancer cells. Anal Bioanal Chem 386:2055–2062

Palomino JC (2005) Nonconventional and new methods in the diagnosis of tuberculosis: feasibility and applicability in the field. Eur Respir J 26:339–350

Park SJ, Taton TA, Mirkin CA (2002) Array-based electrical detection of DNA with nanoparticle probes. Science 295:1503–1506

Patel MK, Ali MA, Srivastav S, Agrawal VV, Ansari SG, Malhotra BD (2013a) Magnesium oxide grafted carbon nanotubes based impedimetric genosensor for biomedical application. Biosens Bioelectron 50:406–413

Patel MK, Ali MA, Zafaryab M, Agrawal VV, Rizvi M, Ansari ZA, Ansari SG, Malhotra BD (2013b) Biocompatible nanostructured magnesium oxide-chitosan platform for genosensing application. Biosens Bioelectron 45:181–188

Patel MK, Ali MA, Krishnan S, Agrawal VV, Al Kheraif AAA, Fouad H, Ansari ZA, Ansari SG, Malhotra BD (2015) A label-free photoluminescence genosensor using nanostructured magnesium oxide for cholera detection. Sci Rep 5:17384

Peter C, Meusel M, Grawe F, Katerkamp A, Cammann K, Borchers T (2001) Optical DNA-sensor chip for real time detection of hybridization events. Fresenius J Anal Chem 371:120–127

Peterlinz KA, Georgiadis RM, Herne TM, Tarlov MJ (1997) Observation of hybridization and dehybridization of thiol-tethered DNA using two-color surface plasmon resonance spectroscopy. J Am Chem Soc 119:3401–3402

Phunpae P, Chanwong S, Tayapiwatana C, Apiratmateekul N, Makeudom A, Kasinrerk W (2014) Rapid diagnosis of tuberculosis by identification of Antigen85 in mycobacterial culture system. Diagn Microbiol Infect Dis 78:242–248

Prabhakar N, Arora K, Arya SK, Solanki PR, Iwamoto M, Singh H, Malhotra BD (2008) Nucleic acid sensor for M. tuberculosis detection based on surface plasmon resonance. Analyst 133:1587–1592

Qian XM, Nie SM (2008) Single-molecule and single-nanoparticle SERS: from fundamental mechanisms to biomedical applications. Chem Soc Rev 37:912–920

Qian XM, Peng XH, Ansari DO, Yin-Goen Q, Chen GZ, Shin DM, Yang L, Young AN, Wang MD, Nie S (2008) In vivo tumor targeting and spectroscopic detection with surface-enhanced Raman nanoparticle tags. Nat Biotechnol 26:83–90

Quinn TC, Wawer MJ, Sewankambo N, Serwadda D, Li C, Wabwire-Mangen F, Meehan MO, Lutalo T, Gray RH (2000) Viral load and heterosexual transmission of human immunodeficiency virus type 1. Rakai Project Study Group. N Engl J Med 342:921–929

Ramsay G (1998) DNA chip: state-of-the art. Nat Biotechnol 16:40–44

Riedel T, Surman F, Hageneder S, Pop-Georgievski O, Noehammer C, Hofner M, Brynda E, Rodriguez-Emmenegger C, Dostálek J (2016) Hepatitis B plasmonic biosensor for the analysis of clinical serum samples. Biosens Bioelectron 85:272–279

Rubtsova M, Presnova DP, Vladimir K, Vitaly G, Alexey E (2017) Biosensor based on a nanowire field-effect transistor for the determination of prostate specific antigen. Proc Technol 27:234–235

Shafiee H, Jahangir M, Inci F, Wang S, Willenbrecht RB, Giguel FF, Tsibris AM, Kuritzkes DR, Demirci U (2013) Acute on-chip HIV detection through label-free electrical sensing of viral nano-lysate. Small 9:2553–2563

Shamsipur M, Nasirian V, Mansouri K, Barati A, Veisi-Raygani A, Kashanian S (2017) A highly sensitive quantum dots-DNA nanobiosensor based on fluorescence resonance energy transfer for rapid detection of nanomolar amounts of human papillomavirus. J Pharm Biomed Anal 136:140–147

Shang-Chen HCW, Chang X, Lu C, Wei C, Lin C, Lai HC, Lin CW (2012) Rapid identification of Mycobacterium tuberculosis infection by a new array format-based surface plasmon resonance method. Nanoscale Res Lett 7:180

Silvestri D, Sonato A, Ruffato G, Meneghello A, Antognoli A, Cretaio E, Dettin M, Zamuner A, Casarin E, Zacco G, Romanato F, Morpurgo M (2015) A peptide nucleic acid label-free biosensor for Mycobacterium tuberculosis DNA detection via azimuthally controlled grating-coupled SPR. Anal Methods 7:4173–4180

Singh R, Dhand C, Sumana G, Verma R, Sood S, Gupta RK, Malhotra BD (2010a) Polyaniline/carbon nanotubes platform for sexually transmitted disease detection. J Mol Recognit 23:472–479

Singh R, Sumana G, Verma R, Sood S, Sood KN, Gupta RK, Malhotra BD (2010b) Fabrication of Neisseria gonorrhoeae biosensor based on chitosan-MWCNT platform. Thin Solid Films 519:1135–1140

Solanki PR, Patel MK, Kaushik A, Pandey MK, Kotnala RK, Malhotra B (2011) Sol-Gel derived nanostructured metal oxide platform for bacterial detection. Electroanalysis 23:2699–2708

Stern E, Vacic A, Rajan NK, Criscione JM, Park J, Ilic BR, Mooney DJ, Reed MA, Fahmy TM (2010) Label-free biomarker detection from whole blood. Nat Nanotechnol 5:138–142

Stougaard M, Lohmann JS, Mancino A, Celik S, Andersen FF, Koch J, Knudsen BR (2009) Single-molecule detection of human topoisomerase I cleavage-ligation activity. ACS Nano 3:223–233

Su M, Li S, Dravid VP (2003) Microcantilever resonance-based DNA detection with nanoparticle probes. Appl Phys Lett 82:3562–3564

Tam PD, Hieu NV, Chien ND, Le AT, Tuan MA (2009) DNA sensor development based on multi-wall carbon nanotubes for label-free influenza virus (type A) detection. J Immunol Methods 350:118–124

Tang D, Yuan R, Chai Y, Zhong X, Liu Y, Dai J (2006) Electrochemical detection of hepatitis B surface antigen using colloidal gold nanoparticles modified by a sol–gel network inter-face. Clin Biochem 39:309–314

Taton TA, Mirkin CA, Letsinger RL (2000) Scanometric DNA array detection with nanoparticles probes. Science 289:1757–1760

Touhami A (2015) Biosensors and nanobiosensors: design and applications. In: Seifalian A, de Mel A, Kalaskar DM (eds) Nanomedicine. One Central Press, Altrincham, pp 375–400

Tripathy S, Vanjari SRK, Singh V, Swaminathan S, Singh SG (2016) Electrospun manganese(III) oxide nanofiber based electrochemical DNA-Nanobiosensor for zeptomolar detection of dengue consensus primer. Biosens Bioelectron 90:378–387

Tsai T (2017) Diagnosis of tuberculosis using colorimetric gold nano-particles on a paper-based analytical device. ACS Sens 2:1345–1354

Wang J (1993) Electroanalysis and biosensors. Anal Chem 65:450–453

Wang J (2000) From DNA biosensors to gene chips. Nucleic Acids Res 28:3011–3016

Wang J, Liu G, Polsky R, Merkoci A (2002) Electrochemical stripping detection of DNA hybridization based on cadmium sulfide nanoparticle tags. Electrochem Commun 4:722–726

Wang J, Liu GD, Merkoci A (2003) Electrochemical coding technology for simultaneous detection of multiple DNA targets. J Am Chem Soc 125:3214–3215

Wang Y, Bai X, Wen W, Zhang X, Wang S (2015) Ultarsensitive electrochemical biosensor for HIV gene detection based on graphene stabilized gold nanoclusters with exonuclease amplification. ACS Appl Mater Interfaces 7:18872–18879

Yu KN, Lee SM, Han JY, Park H, Woo MA, Noh MS, Hwang SK, Kwon JT, Jin H, Kim YK, Hergenrother PJ, Jeong DH, Lee YS, Cho MH (2007) Multiplex targeting, tracking, and imaging of apoptosis by fluorescent surface enhanced Raman spectroscopic dots. Bioconjug Chem 18:1155–1162

Yuna M, Myunga NV, Vasqueza RP, Wangb J, Monbouquette H (2003) Nanowire growth for sensor arrays. SPIE Proc 5220:1–8

Zhang C, Hu J (2010) Single quantum dot-based nanosensor for multiple DNA detection. Anal Chem 82:1921–1927

Zhang CY, Yeh HC, Kuroki MT, Wang TH (2005) Single-quantum-dot-based DNA nanosensor. Nat Mater 4:826–831

Zhao XJ, Dytioco RT, Tan WH (2003) Ultrasensitive DNA detection using highly fluorescent bioconjugated nanoparticles. J Am Chem Soc 125:11474–11475

Zheng L, Jia L, Li B, Situ B, Liu Q, Wang Q, Gan N (2012) A sandwich HIV p24 amperometric immunosensor based on a direct gold electroplating-modified electrode. Molecules 17:5988–6000

Zhou J, Wang QX, Zhang CY (2013) Liposome–quantum dot complexes enable multiplexed detection of attomolar DNAs without target amplification. J Am Chem Soc 135:2056–2059

Zhu NN, Zhang AP, Wang QJ, He PG, Fang YZ (2004) Lead sulfide nanoparticle as oligonucleotides labels for electrochemical stripping detection of DNA hybridization. Electroanalysis 16:577–582

Zhu H, Lu F, Wu X-C, Zhu J-J (2015) An upconversion fluorescent resonant energy transfer biosensor for hepatitis B virus (HBV) DNA hybridization detection. Analyst 140:7622–7628

Genetically Encoded Nanobiosensors for Nutrients and Their Applications

6

Rahila Nazir, Mohd. Mohsin, Manju Nehra, and Tariq Omar Siddiqi

Abstract

Nanotechnology has an intense impact on the improvement of a novel class of biosensors identified as nanobiosensors. Nanobiosensors are broadly used for molecular recognition of biomarkers linked with a diagnosis of diseases. The use of novel nanomaterials in biosensing has inclined biosensing investigation. This chapter sums up the advances in nanosensors for the nutrients, such as, amino acids, sugars, metabolic precursors, and signaling molecules which will help to explain the complex roles of these small molecules in science. In view of the mechanism of genetically encoded sensors, an assortment of novel chemical sensors has been produced for nutrients. Such genetically encoded sensors aid us to visualize the metabolite flux within cells, through which we can solve numerous physiological problems both in plants and animals. Since these sensors are genetically encoded and therefore can be effortlessly targeted to subcellular organelles to monitor the flux in them. This chapter features the general design, different types, and platform of sensors along with the sensitive and specific genetically encoded sensors for nutrients and their applications.

Keywords

Nanobiosensors · Genetically encoded · Nanotechnology · Nutrients · Metabolites

R. Nazir · T. O. Siddiqi
Department of Botany, Jamia Hamdard, New Delhi, India

M. Mohsin (✉)
Department of Biosciences, Jamia Millia Islamia, New Delhi, India
e-mail: mmohsin1@jmi.ac.in

M. Nehra
Department of Food Science and Technology, Chaudhary Devi Lal University, Sirsa, India

M. Mohsin et al. (eds.), *Nanobiosensors for Agricultural, Medical and Environmental Applications*, https://doi.org/10.1007/978-981-15-8346-9_6

Abbreviations

AA's	Amino acids
FPs	Fluorescent proteins
FRET	Fluorescence resonance energy transfer
GE	Genetically encoded
GFP	Green fluorescent proteins
MS	Mass spectroscopy
NMR	Nuclear magnetic resonance

6.1 Introduction

Nanotechnology developments have led to an increase of nanoscale biosensors that have delicate sensitivity and versatility. The indispensable objective of nanobiosensors is to sense any biological signals connected with a particular infection of a single molecule or cell. They can be coordinated into different innovations, for example, lab-on-a-chip to reassure atomic diagnostics. They have applications which consist of the recognition of microorganisms in numerous samples, metabolite monitoring, and detection of tissue-related diseases like cancer. Their movability makes them supreme for the pathogenesis of cancer and can be used in the laboratory settings as well. The capacity to identify diseases related biomolecules, for example, infection explicit metabolites, nucleic acids, proteins, and cells, for example, coursing cancer cells, is important not just for malady determination in the medical set yet additionally for biomedical study including medication disclosure and improvement. Nanotechnology, with its upgraded affectability and decreased instrumentation size, will quickly improve our current bio indicative limit concerning specificity, speed, and cost. Decrease in sensor size gives extraordinary flexibility to fuse into multiplexed, movable, versatile, and even implantable clinical tools (Sauret et al. 2011). Reconciliation of nanoscale ultrasensitive sensors with other clinical tools will make the way for rising clinical fields, including the purpose of care diagnostics and universal human services frameworks. The biomedical use of nanobiosensors is extensive; besides, the upcoming effect of nanobiosensor frameworks for purpose of diagnostics will be matchless. This innovation will change customary clinical performs by empowering primary determination of constant crippling maladies, hyper-sensitive location of pathogens, and extended haul observing of patients utilizing biologically compatible incorporated clinical instrumentation (Allen et al. 1999). In 1960 biosensors were begun by the pioneers Clark and Lyons, different sorts of biosensors utilized are catalyst constructed, tissue centered, insusceptible sensors, DNA biosensors, and electrified biosensors. The main catalyst constructed sensor remained accounted for by Updike and Hicks in 1967. Chemical biosensors have been imagined on incapacitated systems, for instance, the suction of synthetic concoctions by van der Waals powers, ionic holding, or covalent holding. The tissues for tissue-based sensors rise out of plant and animal sources. The analyte of intrigue can be an inhibitor or a substrate of these techniques. Tissue-based sensor was developed by Richnitz for amino acid arginine. Organelle based sensor was also made for chloroplast and mitochondria. For such sort of biosensors, the dependability was extraordinary, however the identification period was extensive, and the explicitness was diminished (Touhami 2014). Immunosensors were set up in transit that antibodies have a high favoritism towards their different antigens, for instance, the antibodies unequivocally bind to microbes or harms or interface with fragments of the host's immune system. Magnetic biosensors, scaled-down biosensors recognizing the magnetic small gage and nanoparticles in microfluidic channels exploiting the magnetoresistance influence have extraordinary latent as far as sensitiveness and magnitude (Hussain et al. 2014). Calorimetric and thermal biosensors were created by engrossing biosensor elements, as referenced earlier into a physical transformer. Green fluorescent protein and the ensuing auto fluorescent protein (AFP) variations and hereditary combination columnists have supported the advancement of hereditarily encoded biosensors. This sort of biosensor is easy to understand, simple to design, control, and move into cells. Single-chain FRET-based biosensor is another model. They comprise of a couple of AFPs, which can move fluorescence reverberation vitality between them when united close. Various techniques might be utilized to manage variations in FRET indications dependent on force, proportion, or time of AFPs (Medintz 2006).

6.2 Genetically Encoded Biosensors and Types (Fig. 6.1)

6.2.1 Fluorescence-Based Biosensors

Biosensors based on fluorescence are imaging operators for use in malignant growth and drug revelation. They have endowed bits of knowledge and guideline of substances at the cellular level. Genetically encoded FRET biosensors undertake an imperative job. Fluorescent biosensors are nanoscopic platforms onto which one or a few fluorescent tests are mounted, genetically, and enzymatically through a receptor (Shaner et al. 2005). The sensory receptor recognizes a specific analyte or aim, accordingly transforming a fluorescent sign which can be quickly identified and estimated. Metabolites, protein biomarkers, and particles can be detected by fluorescent biosensors with unexpected affectability and can also account for the proximity, development, and position of the goal in multifaceted courses of action (Sadanandom and Napier 2010). They have been applied in challenging excellence articulation, protein restriction, and reworking in fields, for instance, signaling, interpretation, cell cycle, and cell death. Sign of joint pain, inflammatory ailments, cardiovascular, and neurodegenerative ailments, viral contamination, malignancy, and metastasis is done utilizing these sensors (Okumoto et al. 2011). Fluorescent biosensors are used in medicate revelation programs for the unmistakable evidence of prescriptions by high throughput, high substance screening, for a post-screening

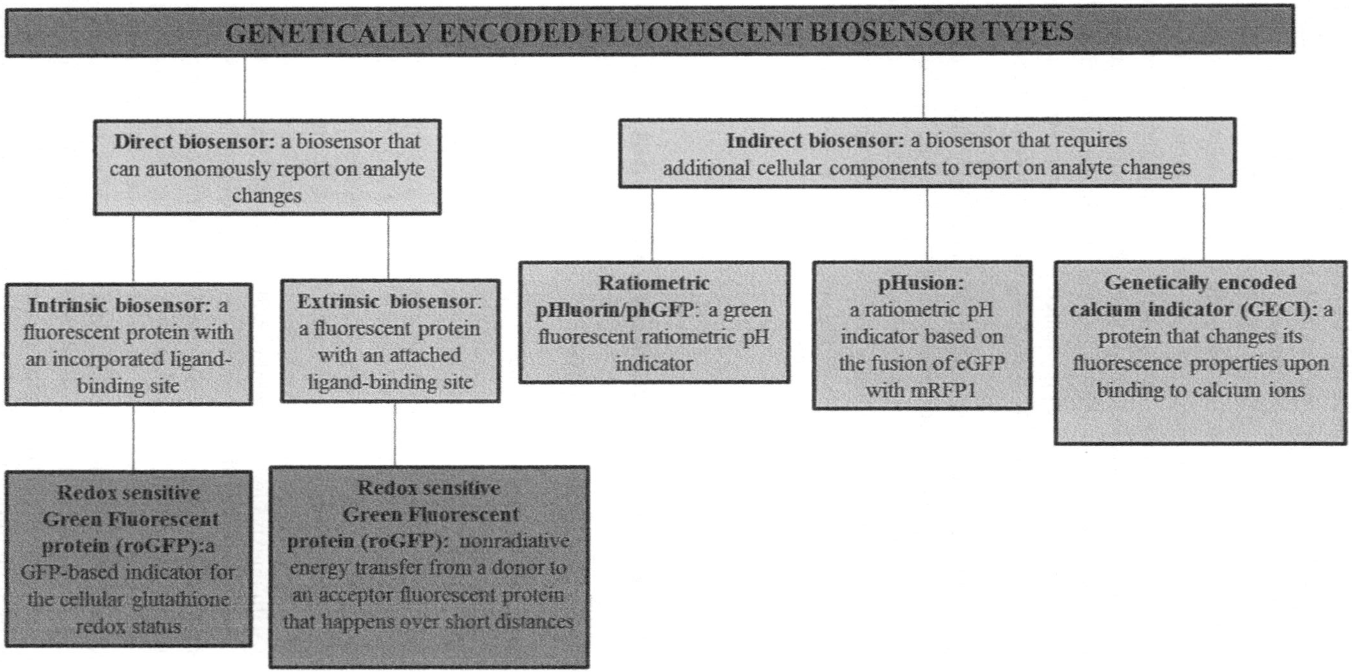

Fig. 6.1 Flowchart diagram showing different types of genetically encoded fluorescent biosensor types

assessment of hits and improvement of leads. For preclinical assessment and clinical authentication of therapeutic potential, bio dissemination, and pharmacokinetics of drugs, these tools are considered very potent (Samodelov et al. 2016). In the discovery of biomarkers in atomic and clinical diagnostics for checking ailment program and response to therapeutics, intravital imaging, and medical procedures, fluorescent biosensors are successfully employed. Genetically encoded FRET-based biosensor for the location of the Bcr-Abl kinase measure was developed on malignant growth patient cells to survey Bcr-Abl kinase action and to set up an interrelation with the infection status in interminable myeloid leukemia. This test was additionally applied to accomplish reaction to usage, and to guard the beginning of drug-resistant cells, permitting expectation for elective therapeutics.

In Metabolic Engineering

Natural concerns and absence of maintainability of oil inferred substances are step by step cautioning the requirement for the improvement of microscopic cell production lines for the consolidation of synthetic concoctions. Experts see metabolic designing as the endowing invention for cost-effective bio-economy. They have additionally imagined that a considerable part of fuels, commodity compounds, and drugs will be created from sustainable feedstocks by using microorganisms as a substitute rather than reliant on oil refining and extraction from plants (Sato et al. 2014). The high limit with respect to decent variety age furthermore requires proficient selection tactics to select the persons carrying the ideal phenotype. The previous strategies were spectroscopy-based enzymatic examination nevertheless they had limited output. To dodge this snag genetically encoded biosensors that permit in vivo observing of cellular metabolism were created which open the possibility for high amount selection and high-quality developing fluorescence-activated cell sorting (FACS) and cell endurance, separately. FRET sensors involved a couple of donor and acceptor fluorophores, and a ligand-restricting peptide was inserted among the two. At the point when it was limited by a ligand of intrigue, the peptide practiced a conformational change in this manner a FRET change. Despite the fact that they had high symmetry, goals, and ease of development, FRET sensors were only organized to account for the bounty of metabolites concerned and could not apply downstream guideline to the indication (Schulte et al. 2006). Transcription aspects are usual sensory proteins that progressed to control gene expression in reaction to variations in the surroundings for high amount of selection. It is practiced by hacking into the host transcription framework and utilizing a manufactured condition explicit promoter to drive the expression of a reporter gene. These show poor symmetry and foundation commotion. The third class of biosensors involves riboswitches; the regulatory domain of an mRNA that can specifically tie to a ligand and immediately vary its own structure, therefore directing interpretation of its encoded protein (Shaner et al. 2007). Rather than TF based biosensors, they are similarly quicker as the RNA has just been interpreted, additionally, they don't depend on protein-protein or protein–metabolite cooperation's. In current years ribosomes have been extensively designed in bacterial frameworks.

6.2.2 Imaging of Ions and Metabolites in Intact Tissue and Organs

Estimating ion and metabolite amount in tissues needs the assurance of fluorescence powers from a few cell layers, providing a substantial practical task for the application of nanosensors. This intervention is characteristically mediated by shielding properties from overlying cells, or by absorption of donor fluorophore emission by acceptor fluorophores from supplementary tissue layers (Vermeer and Munnik 2013). One possible explanation for this might be the restraint of signals to a section of the image by leading the sensors to nuclei (Schuster et al. 2014). On the other hand, the usage of cell layer explicit promoters may aid to progress recognition in undamaged tissues and structures, though at the price of having to generate transgenic lines with cell type-specific promoters. Advanced imaging innovation can additionally develop 4-Dimensional imaging of FRET variations in cells and tissues. The possibilities and entanglements of 4-Dimensional imaging have been looked into broadly (Swarup et al. 2008). Since non-radiative exciton move adjusts the time properties of the FRET donor, it is conceivable to decide the fluorescence time of the joined donor/acceptor emission by fluorescence lifetime imaging microscopy. Investigation of homoFRET, which needs imaging innovation for deciding anisotropic decay, might be an alternative. Nipkow spinning disk, multiphoton fluorescence microscopy, and deconvolution-based methodologies can be utilized to get pictures with higher 3-D goals, an angle particularly significant while imaging in soft tissue or organs of multicellular organisms (Sozzani et al. 2014).

6.3 Sensors for Nutrients

6.3.1 Inorganic Phosphate

Plant life attains and integrates phosphorous in the form of inorganic phosphate (Pi), which is obligatory and used for vital cellular progressions such as energy transfer reactions, signal transmission, and enzyme actions. cpFLIPPi is another age FRET-based biosensor that recognizes Pi and comprises of a cyanobacterial Pi-restricting protein intertwined amid eCFP as FRET donor and cpVenus as FRET acceptor (Vermeer et al. 2009). cpFLIPPi was newly established and utilized in plants to sense fluctuations in cytoplasmic Pi concentrations in root epidermal cells in reaction to Pi starvation and replacement. In addition to this, a plastid-targeted variety of the cpFLIPPi sensor was used to evaluate the part of plastid Pi transporter PHT in Pi transport (Simon et al. 2014).

6.3.2 Zinc

High-affinity FRET developed Zn^{2+} sensors (e.g., eCALWY-1) that comprise two metal-binding domains, ATOX1 and WD4, connected through a flexible linker and flanked by Cerulean as FRET donor and Citrine as a FRET acceptor. Binding of Zn^{2}

+in the middle of the two metal-binding domains grounds a fall in energy transfer that exposes Zn^{2+}absorptions. Using eCALWY-1 and modified forms in Arabidopsis root cells, Lanquar et al. stated on cytoplasmic-free Zn^{2+}concentrations in roots provided with various exogenous Zn^{2+} concentrations using the Root Chip. Experiments associating a FRET-based biosensor with the perfusion switch achieved by the Root Chip validate the link of low-and high-affinity take-up frameworks just as the discharge of internal stores of Zn^{2+}controlling Zn^{2+} homeostasis in living cells (Vinkenborg et al. 2009).

6.3.3 Glutamate

In addition, being a metabolic intermediate and significant artificial precursor, glutamate is the main excitatory amino acid neurotransmitter in the central nervous system of vertebrates. Understanding the dimensions of glutamate can clarify complex neurodegenerative sicknesses, for instance, Parkinson's, and help the consequent medication plan. To screen together intracellular and cell surface glutamate levels, the *E. coli* glutamate/aspartate-restricting protein YBEJ was altered over to a genetically encoded nanosensor by sandwiching in an advanced cyan fluorescent protein (eCFP) donor and an enhanced yellow fluorescent protein (Venus) acceptor. To sense the extensive concentration range, a selection of mutational variations were introduced into the binding pocket of the protein, to generate glutamate sensors whose affinities alternated from high nanomolar concentrations (Schwarzlander et al. 2008).

Extracellular glutamate release was sharply focused by attaching to the plasmid-borne gene both a leader sequence, coordinating the nanosensor to the secretory pathway, and a transmembrane space, which assisted to contend the developed protein to the extracellular plasma film. Discoveries from utilizing these nanosensors in rodent hippocampal cells propose that consistent state glutamate intensities are determined primarily by metabolic modification, as contrasting to take up, and glutamate aggregates at the cell surface formerly discharge by depolarization; all of which function to give understanding into its unpredictable job.

Glucose

In overall, the blood sugar intensities are precisely retained by the liver, which dissipates high blood glucose levels and, on the other hand, discharges deposited glucose when levels drop. Throughout the last step of gluconeogenesis, glucose is formed inside the endoplasmic reticulum (ER) but the direction that glucose follows before circulatory release remains indistinct. Glucose should be moreover first directed out into the cytosol or distributed legitimately over the cell by a vesicular pathway. Glucose nanosensors with nanomolar (nm) and micromolar (mm) affinities were examined in an exceedingly hepatic cell line, the sensors were targeted to the cytosol or the ER lumen by modification of the gene with a localization sequence. Assessment of the sensor terminates in these subcellular compartments found that the glucose levels and metabolic kinetics inside the ER were vague from cytosolic levels, an outcome which may be elucidated by quick and bi-directional glucose transport over the ER membrane. These outcomes led to the knowledge that ER glucose transfer from hepatocytes happens through the cytosol and is catalyzed by a still to be defined set of transport proteins (Wiren et al. 2000).

Ion Sensors

FRET-based discovery has developed a prevailing tool for quantifiable amounts of numerous analytes such as H^+, metal ions, and glucose in environmental, industrial, therapeutic, and living applications as of its sensitivity, specificity, and real-time monitoring with quick response time (Schumacher et al. 2020). Dibyendu et al. discovered a tactic for the sensing of ions by determining the concentration of comparing salts KCl, NaCl, $MgCl_2$, $CaCl_2$, $FeCl_3$, $FeSO_4$, $AlCl_3$ in water, in view of FRET between two laser colors Acriflavine and Rhodamine B. The principle of the planned sensor is based on the modification of FRET efficiency among the dyes in the occurrence of diverse ions K^+, Na^+, Mg^{2+}, Ca^{2+}, Fe^{2+}, Fe^{3+}, Al^{3+}. Nano dimensional clay platelet laponite was used to progress the efficiency of the sensor. Transition metal ions play an important part in biology as nutritional microelements as well as dynamic ligands in proteins and small molecules. The monitoring of deadly metal ions in water ecologies is a significant matter since these pollutants can have unembellished outcomes on human well-being and the environment (Shih et al. 2015). Lead and mercury are two of the utmost toxic metallic contaminants; for instance, lead can cause kidney malfunction and hinder brain growth and mercury can harm the brain, heart, and kidneys. Mercury pollution is a universal concern and the important source of human experience stems from polluted natural waters. Mercury experiences long go transport in the earth between different media, for instance, air, soil, and water by testament as of anthropogenic releases. The atmospheric oxidation of mercury vapor to water-soluble Hg^{2+} ions and its consequent metabolism by aquatic microorganisms produces methyl mercury, a strong neurotoxin associated with several cognitive and motion ailments. Besides these lines, attaining new mercury detection methods that are economically savvy, quick, easy, and relevant to the natural and organic milieus is an important objective. Liu et al. detailed a FRET-based ratiometric sensor for the location of Hg^{2+} ion. Silica nanoparticles were noticeable with a hydrophobic fluorescent nitrobenzoxadiazole dye which goes about as a FRET donor. Rhodamine was then covalently connected to the surface of the silica particles which acts as acceptor. Nanoparticles are observable to Hg^{2+} in water. FRET-based system with control over the location of both donor and acceptor and their parting distance within the nanoparticles has been recognized for ratiometric sensing of Hg^{2+} in water. The standard of detecting was based on FRET from pyrene excimer emissions to ring-opened rhodamine absorption upon complexation of the Hg^{2+} ion. Chao et al. detailed a FRET-based ratiometric discovery outline for mercury particles in water with polymeric particles as platforms. FRET-based sensors can possibly make time-subordinate focus or action maps of ions, small ligands, or macromolecules in living cells. So as to address the difficulty of multidimensional perception, the vibrant range and reaction energy of the biosensors are elementary features, since they straightforwardly influence the sensor's spatial and temporal resolution. A time-resolved microfluidic flow cytometer prepared for describing the FRET-based commanding reaction of metal-ion sensors in mammalian cells has remained structured. The tool can be used to examine the cellular heterogeneity of Zn^{2+} and Ca^{2+} sensor FRET reaction signals. Very thoroughly 30 overlay dissimilarity between the extracellular and intracellular sensors has been accounted for FRET-based Cd^{2+}symbolcomprising a Cd^{2+}limiting protein got from *Pseudomonas putida* as the Cd^{2+}sensing key has been accounted for live cell detecting of Cd^{2+} (Okumoto et al. 2011).

Hard Water Sensor

The inorganic content of hard water is very high comparison to soft water. Though hard water is not hazardous to one's health normally but can cause severe difficulties in industrial localities, where water hardness must be observed to evade failures of the high-priced tools that hold water. The firmness of water is determined by the concentration of multivalent cations in water. The most common cations found in hard water comprise Ca^{2+} and Mg^{2+}. The impermanent hardness in water includes the presence of dissolved carbonate minerals ($CaCO_3$ and $MgCO_3$), which can be concentrated either by boiling the water or by the addition of lime (calcium hydroxide). In contrast, the dissolved chloride minerals ($CaCl_2$ and $MgCl_2$) cause the lifelong hardness of water that cannot be separated simply because it becomes more soluble as the temperature rises. In that sense, it is very important to distinguish the permanent hardness of water before use. The association of absorption or fluorescence spectroscopy for water valuation has gotten unambiguous consideration. Sweetser and Bricker were the scientists who used the spectroscopic assessments to choose the concentration of calcium and magnesium particles in water. Ion chromatography (IC) is another usual policy for the investigations of anions and cations in the aqueous solution. The FRET phenomenon is a very effective tool for the designing of hard water sensors. In this case, the result of Mg^{2+} or Ca^{2+} or both on the FRET efficiency between two fluorophores, Acriflavine (Acf) and rhodamine B (RhB) in the presence of nano clay sheet laponite has been measured. The analysis showed that FRET efficiency drops with increasing ion Mg^{2+} or Ca^{2+} or both concentration. This is as both the dyes Acf and RhB used were cationic in nature. The inclusion of cations increases the separation between them, causing a decrease in FRET efficiency. Nano clay platelet laponite was used to increase sensing efficacy. It has also been proved that with proper calibration, this sensor can be used to sense water hardness with an adequate resolution between soft water salt concentration less than 0.06 mg/ml, moderately hard water salt concentration more than 0.06 mg/ml and less than 0.12 mg/ml and the hard water salt concentration above 0.12 mg/ml (Sieberer et al. 2012).

FRET-Based pH Sensors

Detecting of pH is one of the important methods which are crucial in innumerable fields of application reaching from agriculture and environment to industry, medicine, and food. In medical science, abnormal pH values inside the cell specify inappropriate cell function, growth, and division. It is also useful to diagnose some common diseases like cancer and Alzheimer's. For the sensing of pH, there are two very prominent practices, in particular, Optical chemical sensors, similarly named optrodes, and FRET-based pH sensors. On the version of optrodes, the alteration in absorbance or fluorescence force of the pH-sensitive dyes shows a modification in the pH of the environment (Shen et al. 2013). On the other hand, FRET-based pH sensors are verified by the ratiometric variations of the dye fluorescence of both donor and acceptor with the pH of the environment. Optrodes exploit pH indicator dyes, weak organic acids, or bases, with diverse optical properties associated with their protonated and deprotonated forms. The absorption or fluorescence properties of these dyes are reformed with a variation of pH of the environment (Schumacher 2014).

6.4 Biological Discoveries Made with Biosensors

Some of them are explained below:

6.4.1 Calcium Imaging in Guard Cells

Calcium imaging in plants utilizing FRET-based yellow cameleons has been directed through work done in guard cells. In guard cells, yellow cameleons show spontaneous cytoplasmic calcium motions, just as calcium motions that are triggered or attuned by means of outer consumptions of calcium, ABA, MeJA, H_2O_2, sorbitol, yeast elicitor (YEL), chitosan, thiocyanate, chitin, and changes in CO_2. Calcium transients in guard cells can be artificially attained through another perfusions with hyperpolarizing and depolarizing buffers (Suzuki et al. 2014). Through such an outline, the perfect calcium strategy for the stomatal closure was defined by three 5-min transients at times of 10 min (Sieberer et al. 2012). A test in guard cell calcium-flagging examination has been characteristic guard cell plasma membrane calcium penetrable (ICa) channel encoding genes that are initiated by ABA, H_2O_2, MeJA, YEL, and chitosan. Studies using yellow cameleons have added to the characterization of mutants intervening such as calcium reactions by means of ICa channels (Tang et al. 2011).

Calcium Imaging in Pollen Tubes and During Fertilization

Calcium imaging in pollen tubes was initially predictable with yellow cameleon. In *Lilium longiflorum* and *Nicotiana tabacum*, investigators described a calcium incline along with tip-focused oscillations (Thestrup et al. 2014). During in vitro *Arabidopsis* pollen tube growth, the calcium porous channel cyclic nucleotide-gated channel CNGC18, ROS-producing NADPH oxidases AtRBOHH and AtRBOHJ, and D-serine affected tip-focused calcium oscillations (Thor and Peiter 2014). The pharmacological and genetic indication also specifies that glutamate receptor-like channels play a part in the self-incompatibility response (Steinhorst and Kudla 2013).

6.4.2 Visualization of Reactive Oxygen Species (ROS) and Redox Changes

ROS are reactive systems of sub-atomic oxygen and are designed as fatal outcomes in metabolic responses. However, they additionally go about as indicating atoms to facilitate metabolic, expansion, and developmental events. ROS levels are important for life and should be kept over a cytostatic so far under a cytotoxic level to allow proper redox biology. ROS intensities are accomplished through the intensive action of subcellular compartmentalized ROS creating and ROS scavenging constituents to retain up a perfect cell redox state. ROS and redox biosensors mostly comprise intrinsic probes, roGFPs, and rxFPs, to monitor the glutathione redox state and extrinsic biosensors for the NAD+/NADH ratio, H_2O_2 (HyPer-family and modified roGFPs), and other ROS. In plants, roGFPs have been used mostly to regulate the subcellular glutathione redox potential and to quantify the glutathione redox state in mutants that are involved in glutathione biosynthesis (Tian et al. 2009). Experiments using roGFPs were newly completed to examine the effects of abiotic stress on organelle redox dynamics. Long-term treatments indicate increased glutathione oxidation in reaction to numerous stresses and some degree of organelle specificity that is reliant on the stress. Short-term analyses of (H_2O_2) changes in *Arabidopsis* guard cells and roots (Stepanova et al. 2007). Studies indicate that H_2O_2 scavenging in peroxisomes might best emulated via artificial calcium lavations that may generate the activation of catalases (Walia et al. 2018)

6.4.3 Hormones

TwoFRET-based biosensors for the phytohormone ABA have been developed that report on ABA dynamics in response to exogenous ABA or challenge with stress conditions. ABACUS1 (abscisic acid concentration and uptake sensor) was engineered by linking an ABA sensory module (PYL1 fused to a highly truncated ABA interaction domain of ABI1) to edCeruleanas FRET donor and edCitrineas FRET acceptor. The ABACUS1 biosensor was used to identify reversible and dose-dependent ABA accumulation following pulsed ABA treatments in roots developing in RootChip16, representing that the ABA end rate is cell specific and improved by ABA (Swarup et al. 2008). A relative yet specific ABA biosensor, ABAleon2.1, was designed by combining an ABA sensory module (PYR1 connected to synergist area of the ABI1) to mTurquoise as FRET donor and cpVenus as FRET acceptor ABAleon2.1 gave an account of endogenous ABA concentrations in response to abiotic stresses and was utilized to follow the significant distance translocation of ABA among root, hypocotyl, and shoot tissues because of exogenous utilization of ABA. A recent FRET-based biosensor, gibberellin observation sensor1 (GPS1), recognizes Gibberellic acid (GA) and was designed by intertwining a GA sensory module with edCerulean as FRET donor and edAphrodite as FRET acceptor (Silverstone et al. 2007). GPS1 reacts to nanomolar concentrations of bioactive GAs and displays moderate obvious reversibility in vitro (Ulmasov et al. 1997).

Hence, GPS1 communicated in vivo can report an increase in GA yet does not give an account of GA consumption. Expression of GPS1 from ap16 promoter, which goes around silencing observed previously, licenses the discovery of an inclination of GA in dark-grown hypocotyls of wild-type and light-signaling mutants (Uslu and Grossmann 2016). GPS1 reports higher GA levels in bigger cells of the dark-grown hypocotyl as compared with small cells close to the apical hook, and apparent GA accumulation in the dark is diminished in a photochromic-interacting factor quadruple mutant. GPS1 additionally reports an endogenous GA gradient in primary root tips, which is reflected by an accumulation of exogenous GA4. Accordingly, GA designing could be accomplished in roots freely of patterns of GA biosynthesis.

Applications of Nanobiosensors

The meaning and depiction of the idea of the process of nanobiosensors do not leave any room for their applications as they are highly multipurpose and multifunctional. The following are some main applications:

- Medical and analytical applications
- Environmental applications
- In food manufacturing, monitoring, originality, standard, and safety
- In fermentation processes
- In plant biology

6.5 Conclusion and Future Perspective

Genetically encoded nanobiosensors comprise a number of profits including sensor design, a sensor concentration-independent outcome signal, and precise subcellular aiming. In addition, genetic encoding permits suitable distribution amongst cell biologists, as an increasing quantity of FRET sensors have become accessible through depositories such as Add Gene. The expansion of responsive nanobiosensors has confirmed to be stimulating in many cases but in current years has advanced both from rational proposal tactics and from directed evolution-like methods. The benefits of expending genetically encoded sensors are actively noticeable since they prevent the difficulties of intracellular delivery and directing that are compulsory for other groups of sensors. Furthermore, these sensors not only permitting the measurement of steady-state concentrations but, more conspicuously, they permit monitoring of in vivo kinetics or flux, an important fact that is frequently overlooked. This bids a special means to study which protein(s) switch or are involved with in vivo flux.

6.6 Summary

Genetically encoded nanobiosensors are promising tools for real-time monitoring of metabolites with cellular or subcellular resolution. Nanobiosensors for nutrients such as sugars, metabolic precursors, AA's, Carbohydrates, vitamins, minerals, and different other metabolites will help to explain the multifarious roles of these small molecules in biology. Since these nanobiosensors are GE they can be targeted to DNA level in plants as well as animals. Different methods, such as fast atom bombardment MS and NMR spectroscopy have been used to determine the level of metabolites in living cells, however, these methods require the disturbance and breakdown of tissues and cells. Nanobiosensors present an opening to monitor real-time dynamics of molecular events in numerous cell and tissue types. By using these nanobiosensors, we can assimilate with cell biological data, high-resolution transcriptional profiling, and computational modeling to shape a combined picture of biological processes. Nanobiosensors based on FPs and FRET (Fluorescence resonance energy transfer) have made an intense impact on Nanotechnology. By using different GFP variants many sensors have been developed. Biosensing has been one of the newest topic fascinating scientific minds since long back. It is because the biological entities are very compound and are directly connected with the existence of a healthy environment. The design of nanobiosensors also has viewed remarkable changes in the recent past. Biosensors for applications as diverse as food quality assessment, environmental monitoring, and diagnosis of clinical and metabolic obstacles have come to the forefront. Nanotechnology has bestowed some highly exhilarating elements for the improvement of sensing phenomenon. The Nanobiosensors have attained significant accomplishment both in the commercial and academic fields and that the need for novel, easy-to-use, home, and reorganized diagnostics is greater than ever. This chapter highlights the different types of genetically encoded nanobiosensors for nutrients and their implicational aspects.

Acknowledgment The first author (RN) is thankful to the Hamdard National Foundation (HNF) Jamia Hamdard, New Delhi for providing the fellowship.

References

Allen GJ, Kwak JM, Chu SP, Llopis J, Tsien RY, Harper JF, Schroeder JI (1999) Cameleon calcium indicator reports cytoplasmic calcium dynamics in *Arabidopsis* guard cells. Plant J 6:735–747

Hussain SA, Dey D, Chakraborty S, Saha J, Roy DA, Chakraborty S, Debnath P, Bhattacharjee D (2014) Fluorescence resonance energy transfer (FRET) sensor. J Spectrosc Dyn 10:16–40

Medintz I (2006) Recent progress in developing FRET-based intracellular sensors for the detection of small molecule nutrients and ligands. Trends Biotechnol 12:539–542

Okumoto S, Desuschle K, Fehr M, Hilpert M, Lager I, Lalonde S, Laager L, Persson J, Schmidt A, Frommer W (2011) Genetically encoded sensors for ions and metabolites. J Soil Sci Plant Nutr 50:7947–7953

Sadanandom A, Napier RM (2010) Biosensors in plants. Curr Opin Plant Biol 13:736–743

Samodelov SL, Beyer HM, Guo X, Augustin M, Jia KP, Baz L, Zurbriggen MD (2016) StrigoQuant: a genetically encoded biosensor for quantifying strigolactone activity and specificity. Sci Adv 11:160–1266

Sato EM, Hijazi H, Bennett MJ, Vissenberg K, Swarup R (2014) New insights into root gravitropic signaling. J Exp Bot 8:2155–2165

Sauret GS, Calder G, Harberd NP (2011) Transient gibberellin application promotes Arabidopsis thaliana hypocotyl cell elongation without maintaining transverse orientation of microtubules on the outer tangential wall of epidermal cells. Plant J 4:628–639

Schulte A, Lorenzen I, Bottcher M, Plieth C (2006) A novel fluorescent pH probe for expression in plants. Plant Methods 6:2–7

Schumacher K (2014) pH in the plant endomembrane system — an import and export business. Curr Opin Plant Biol 22:71–76

Schumacher K, Vafeados D, McCarthy M, Sze H, Wilkins T, Chory J (2020) The Arabidopsis det3 mutant reveals a central role for the vacuolar H+-ATPASE in plant growth and development. Genes Dev 13:3259–3270

Schuster C, Gaillochet C, Medzihradszky A, Busch W, Daum G (2014) A regulatory framework for shoot stem cell control integrating metabolic, transcriptional, and phytohormone signals. Dev Cell 28:438–449

Schwarzlander M, Fricker MD, Muller C, Marty L, Brach T (2008) Confocal imaging of glutathione redox potential in living plant cells. J Microsc 231:299–316

Shaner NC, Steinbach PA, Tsien RY (2005) A guide to choosing fluorescent proteins. Nat Methods 2:905–9 176

Shaner NC, Patterson GH, Davidson MW (2007) Advances in fluorescent protein technology. J Cell Sci 120:4247–4260

Shen J, Zeng Y, Zhuang X, Sun L, Yao X (2013) Organelle pH in the Arabidopsis endomembrane system. Mol Plant 6:1419–1437

Shih HW, DePew CL, Miller ND, Monshausen GB (2015) The cyclic nucleotide-gated channel CNGC14 regulates root gravitropism in *Arabidopsis thaliana*. Curr Biol 25:3119–3125

Sieberer BJ, Chabaud M, Fournier J, Timmers AC, Barker DG (2012) A switch in Ca^2+ spiking signature is concomitant with endosymbiotic microbe entry into cortical root cells of *Medicago truncatula*. Plant J 69:822–830

Silverstone AL, Tseng TS, Swain SM, Dill A, Jeong SY (2007) Functional analysis of SPINDLY in gibberellin signaling in *Arabidopsis*. Plant Physiol 143:987–1000

Simon ML, Platre MP, Assil S, van Wijk R, Chen WY (2014) A multi-colour/multi-affinity marker set to visualize phosphoinositide dynamics in *Arabidopsis*. Plant J 77:322–337

Sozzani R, Busch W, Spalding EP, Benfey PN (2014) Advanced imaging techniques for the study of plant growth and development. Trends Plant Sci 19:304–310

Steinhorst L, Kudla J (2013) Calcium - a central regulator of pollen germination and tube growth. Biochim Biophys Acta 1833:1573–1581

Stepanova AN, Yun J, Likhacheva AV, Alonso JM (2007) Multilevel interactions between ethylene and auxin in *Arabidopsis* roots. Plant Cell 19:2169–2185

Suzuki J, Kanemaru K, Ishii K, Ohkura M, Okubo Y, Iino M (2014) Imaging intra organellar Ca^2+ at subcellular resolution using CEPIA. Nat Commun 5:41–53

Swarup K, Benkova E, Swarup R, Casimiro I, Peret B (2008) The auxin in flux carrier LAX3 promotes lateral root emergence. Nat Cell Biol 10:946–954

Tang S, Wong HC, Wang ZM, Huang Y, Zou J (2011) Design and application of a class of sensors to monitor Ca^2+ dynamics in high Ca^2+ concentration cellular compartments. PNAS 108:16265–16270

Thestrup T, Litzlbauer J, Bartholomaus I, Mues M, Russo L (2014) Optimized ratiometric calcium sensors for functional in vivo imaging of neurons and T lymphocytes. Nat Methods 11:175–182

Thor K, Peiter E (2014) Cytosolic calcium signals elicited by the pathogen-associated molecular pattern flg22 in stomatal guard cells are of an oscillatory nature. New Phytol 204:873–881

Tian L, Hires SA, Mao T, Huber D, Chiappe ME (2009) Imaging neural activity in worms, flies and mice with improved GCaMP calcium indicators. Nat Methods 6:875–881

Touhami A (2014) Biosensors and nanobiosensors: design and applications. Nanomedicine 30:374–403

Ulmasov T, Murfett J, Hagen G, Guilfoyle TJ (1997) Aux/IAA proteins repress expression of reporter genes containing natural and highly active synthetic auxin response elements. Plant Cell 9:1963–1971

Uslu VV, Grossmann G (2016) The biosensor tool box for plant developmental biology. Curr Opin Plant Biol 29:138–147

Vermeer JE, Munnik T (2013) Using genetically encoded fluorescent reporters to image lipid signalling in living plants. Methods Mol Biol 1009:283–289

Vermeer JE, Thole JM, Goedhart J, Nielsen E, Munnik T, Gadella TW (2009) Imaging phosphatidylinositol 4-phosphate dynamics in living plant cells. Plant J 57:356–372

Vinkenborg JL, Nicolson TJ, Bellomo EA, Koay MS, Rutter GA, Merkx M (2009) Genetically encoded FRET sensors to monitor intracellular Zn^2+ homeostasis. Nat Methods 6:737–740

Walia A, Waadt R, Jones A (2018) Genetically encoded biosensors in plants: pathways to discovery. Annu Rev Plant Biol 69:497–524

Wiren N, Gazzarrini S, Gojon A, Frommer WB (2000) The molecular physiology of ammonium uptake and retrieval. Curr Opin Plant Biol 3:254–261

Development of Highly Sensitive Optical Sensors Based on Carbon Nanotube (CNTs)

7

Mohammad Talib, Samrah Manzoor, Priyanka Gulia, Nishant Tripathi, and Prabhash Mishra

Abstract

Photodetectors play an essential role in various fields such as defense, imaging, communication system, sensing, binary switch, environmental monitoring, plume detection, and so on. Recent advancements in device technology lead to the development of highly efficient detectors working in harsh environmental conditions. From past decades numerous materials have been experimented for photodetector applications, however, the introduction of nanomaterials has led to higher photon absorption, better responsivity, and faster switching speed. In particular carbon nanomaterials such as carbon nanotubes provides a high surface to volume ratio, quantum confinement effect, high charge carrier mobility, and ultra-wide absorption spectrum which makes them an ideal candidate for photodetection. In this chapter, we have discussed different photodetectors which have been realized using carbon nanotubes. The results reveal that carbon nanotubes act as an active material for ultra-broadband detection from ultraviolet to terahertz range. Different device architecture and composites were introduced for enhancement in the photodetection properties.

M. Talib · S. Manzoor · P. Gulia
Centre for Nanoscience and Nanotechnology, Jamia Millia Islamia, New Delhi, India

N. Tripathi
Department of Nanoengineering, Samara National Research University, Samara, Russia

P. Mishra (✉)
Centre for Nanoscience and Nanotechnology, Jamia Millia Islamia, New Delhi, India

Department of Nanoengineering, Samara National Research University, Samara, Russia

Center for Photonics and 2D Materials, Moscow Institute of Physics and Technology (MIPT), Dolgoprudny, Russia
e-mail: pmishra@jmi.ac.in

M. Mohsin et al. (eds.), *Nanobiosensors for Agricultural, Medical and Environmental Applications*, https://doi.org/10.1007/978-981-15-8346-9_7

Keywords

CNTs · Optical detector · Sensing mechanism

7.1 Introduction

One-dimensional nanostructures like nanotubes, nanobelts, and nanoribbons have recently attracted the interest of the scientific community due to their unique properties (Yan et al. 2009). These nanostructures provide optical, quantum, surface, and dielectric confinement effects which can alter its bandgap, the density of states, light absorption ability, carrier mobility, and optoelectronic properties (Hu et al. 2012; Talib et al. 2019). Devices fabricated using one-dimensional nanostructures like photodetectors, light-emitting diodes, solar cells, and field-effect transistors have shown better properties and overcome limitations such as dark current effects, outside interference, and requirements of sophisticated instrumentations in comparison to their bulk counterparts (Huang and Banfield 2005; Zheng et al. 2015). Among them, Photodetectors find a unique place in the field of optoelectronic devices. Photodetectors that convert the light signal into electrical signals are an essential element for various fields like optical communication, imaging applications, defense, environmental monitoring, chemical, medical, and security purposes. Nanostructures based photodetectors show wider detection broadband, enhanced photoresponsivity, higher photoconductive gain, high external quantum efficiency, light trapping, good stability, high reliability, low cost, and faster switching speed (Wu et al. 2011). Various 1D materials have been used for fabrications of photodetectors like InAsSb, ZnTe, GaN, In_2Te_3, carbon nanomaterials, ZnS, $Zn_xCd_{1-x}Se$, ZnO, CdS, InSe (Dai et al. 2013; Posada et al. 2012; Svensson et al. 2013; Li et al. 2010; Wang et al. 2012). The optical response of these fabricated photodetectors depends on the crystallinity of material, device architecture, doping, surface adsorption, and dimensions (Zhou et al. 2014).

Carbon nanomaterials based optical sensor detects broadband of electromagnetic radiation from ultraviolet to infrared, now the addition of terahertz range opens a new window for medical imaging, security, and food technology (Fig. 7.1) (He et al. 2015). The way carbon nanotubes interact with electromagnetic radiation is unique due to its novel size, tunable bandgap, quantum confinement, and optoelectronic properties therefore researchers focused on this material for improvement in optical sensor technology (Levitsky and Euler 2003). In fact, the use of nanomaterials and nanodevices provides new form factors, flexibility, tenability, etc. Due to the continuous efforts of the scientific community in designing and fabrication of

Fig. 7.1 CNT detection band (Reprinted with permission from Liu et al. 2018)

devices, many improvements have been seen in the last decades in the carbon nanotubes based optical sensor technology.

Recently low-dimensional carbon materials own novel optoelectronic and electronic properties that are auspicious for diverse THz device applications, such as THz detectors, emitters, and polarizers (Ren et al. 2012). The demand for THz detectors that can deliver real-time imaging for a broad range of industrial applications has impelled research into low cost, flexible THz imaging systems. The bendability of the new THz imager and the prospect of fine-tuning will expand the range of CNT-based devices that could be established in the future. Carbon nanotubes (CNT) based ultra-broadband photodetectors show potential because of CNT absorption spectra which covers the entire ultraviolet to the terahertz range. CNT photodetectors involve a strong electric field continually because of the high binding energy of excitons, asymmetric electrical contacts, and hybrid structures with other materials (Avouris et al. 2008; Bonaccorso et al. 2010; Hartmann et al. 2014).

7.1.1 Carbon Nanotubes Properties for Photodetection

Depending on its symmetry and diameter carbon nanotubes may be semiconducting or metallic. The band diagram for metallic and semiconducting CNTs is shown in Fig. 7.2. For metallic CNTs interband transitions between linear bands are not permissible only transitions between parabolic bands are allowed due to the symmetry although the density of states is constant for broad range of energies. As a result, absorption from UV to NIR is dominated by interband transitions similar to semiconducting nanotubes (Spataru and Leonard 2010; Ostojic et al. 2004).

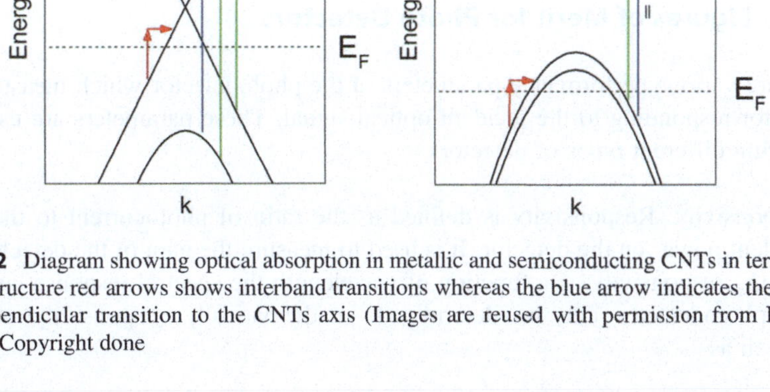

Fig. 7.2 Diagram showing optical absorption in metallic and semiconducting CNTs in terms of its band structure red arrows shows interband transitions whereas the blue arrow indicates the parallel or perpendicular transition to the CNTs axis (Images are reused with permission from He et al. 2015). Copyright done

Table 7.1 Comparative statement of CNTs based photodetector

S. No.	Nanomaterials	Detection Range	Outcomes	Ref.
1.	Tungstite and tungsten oxide decorated MWCNT	365nm Intensity range= 15–83 μW cm^{-2}	Responsivity = 7.4 A W^{-1} Response Time= 400 μs Recovery Time= 500 μs	Majumder et al. (2020)
2.	PbS quantum dots decorated with DWCNTs	433 nm, Intensity = 82 mW cm^{-2}	Responsivity = 30 A W^{-1} Response time = 30 μs	Ka et al. (2020)
3.	SWNTs-Graphene	365 nm, Intensity = 3.9 μW	Photoresponsivity = 204.5 A W^{-1}	Zhanga et al. (2020)
4.	Zinc oxide nanoparticle and SWNTs heterojunction	365 nm, 0.47 mW/cm^2	Recovery Time = 2.7 s Response time = 6.0 s	Choi et al. (2020)
5.	Self-aligned SWNTs	350–1000 nm, Intensity = 50–100 mW	Photoresponsivity = 292.5 mA W^{-1}	Zhou et al. (2019)

The optical properties of finite-length carbon nanotubes depend on its chiral symmetry, doping, and diameter, for example, finite-length carbon nanotubes (n, m) with n-m being a multiple of 3 will have larger values of β and much larger value of α then other length tubes. The photoresponse of the carbon nanotubes UV detectors was explained in terms of π-plasmonic absorption and the photoresponse from visible to NIR region is basically due to the interband transitions in the CNTs. When extended to the mid-IR and terahertz regions, the photoresponse is mainly caused by the plasmon resonance of free carriers in nanotubes (Pichler et al. 1998; Itkis et al. 2002). We included a Table 7.1 comparative survey on present photodetection technology based on CNTs.

7.1.2 Figures of Merit for Photo Detectors

There are various performance parameters of the photodetector which measures its ability for responding to the incident optical signal. These parameters are used for comparing different types of detectors.

1. Responsivity: Responsivity is defined as the ratio of photocurrent to the total incident power on the detector. It is used to measure the gain of the detector. By plotting responsivity as a function of wavelength the spectral response curve of the photodetector can be determined. mathematically responsivity can be written as

$$R_\lambda = \frac{\Delta I}{PS}$$

where ΔI is the photocurrent produced due to illumination of light, P is the light power incident on the detector and S is the surface area of the detector illuminated by the light. Responsivity is measured in AW^{-1}.

2. Quantum efficiency is defined as the ratio of the number of charge carriers generated to the total number of photon incident on the detector. It is a measure of no photons which generate the charge carriers in a detector. Mathematically quantum efficiency defined as

$$QE = hc\frac{R_\lambda}{e\lambda}$$

where e is the electronic charge ($e = 1.6 \times 10^{-19}$ C), λ is the excitation wavelength, h is Planck's constant ($h = 6.626 \times 10^{-34}$ J s), c is the velocity of the light (3×10^{8} m/s), and R_λ is the responsivity of the detector, Quantum efficiency is a dimensionless quantity.

3. The ability of the detector to measure weak optical signal is defined in terms of detectivity, higher is the detectivity better is the performance of the detector. It is measured in terms of m-(Hz)/w. Mathematically detectivity is calculated using the formula,

$$D = \frac{R_\lambda {A_d}^{1/2}}{(2eI_d)^{1/2}}$$

where R_λ is the responsivity of the detector, 'A_d' is an effective area, 'e' is the charge on the electron, and 'I_d' is the dark current.

4. Noise equivalent power of the detector is defined as the ability to measure minimum detectable illumination power. It is a measure of the power which gives a unit signal to noise ratio at 1 Hz bandwidth. Mathematically it can be calculated using the formula,

$$\mathrm{NEP} = \mathrm{PSD}/R_\lambda$$

where PSD is the power spectral density at 1 Hz bandwidth and R is the responsivity of the detector. It is measured in units of W $Hz^{-1/2}$.

5. The sensitivity of the detector is defined as the ratio of change in resistance to the initial resistance of the detector under illumination with light. Mathematically sensitivity defined as

$$\text{Sensitivity} = \Delta R / R_i \times 100\ (\%)$$

where ΔR is the change in resistance of the detector and R_i is the initial resistance.

7.1.3 Carbon Nanotubes Based Optical Detectors

The first report on CNT photoconductivity was observed by Fujiwara et al. (2001). In their experiment, CNT photoconductivity spectra exhibit two peaks at 0.7 and 1.2 eV corresponding to the successive gap of CNTs. The photoconductive response at room temperature and low temperature is recorded and it is found that the response is higher at low temperature than room temperature (Fig. 7.3). Also, temperature-dependent photoresponse was measured at an excitation energy of 0.7 and 1.2 eV.

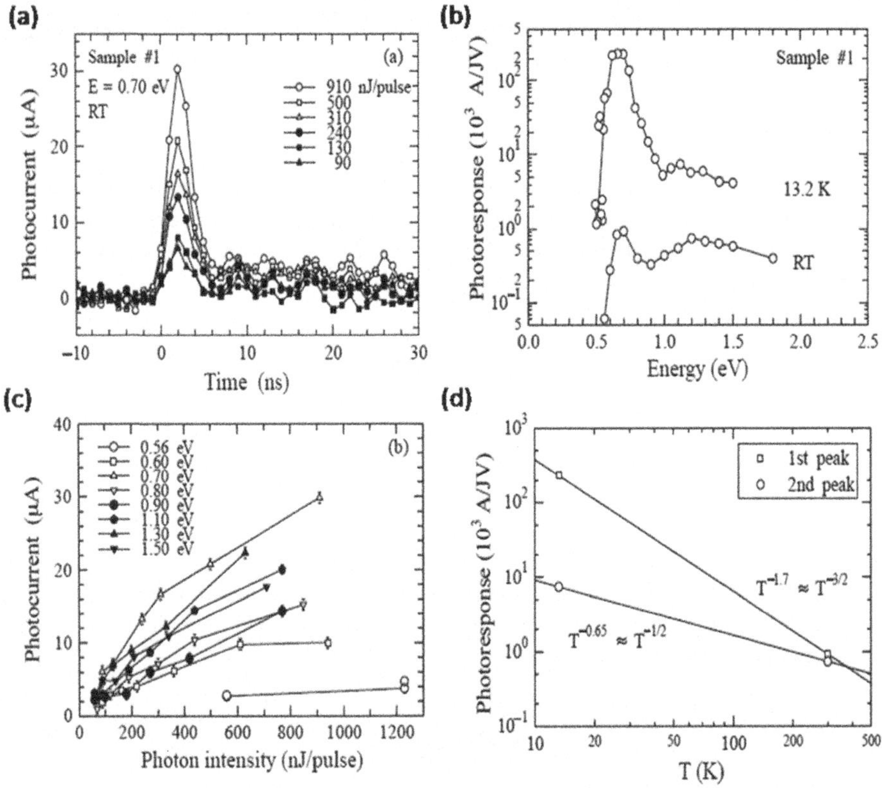

Fig. 7.3 (**a**) Photocurrent versus time graph at room temperature for different light intensities. (**b**) Photoresponse versus excitation energy spectra at 13.2 K and room temperature. (**c**) Photocurrent as a function of light intensity at different bias. (**d**). Temperature-dependent photoresponse at an excitation energy of 0.7 and 1.2 eV. Figure **a**, **b**, **c**, and **d** is reproduced with permission from Fujiwara et al. 2001. Copyright done

Linear relationship between photocurrent and light intensity was observed at different bias voltages. The detailed mechanism related to the unusual temperature dependence was not clearly defined in their report and more experimental studies suggested to completely understand this behavior (Fujiwara et al. 2001). In 2003 ambipolar field-effect transistor was made using a single carbon nanotube channel between the source and the drain electrode. Polarization dependent photocurrent spectra and laser power versus photocurrent spectra were recorded under infrared light illumination (Fig. 7.4) (Freitag et al. 2003). Lu and Panchapakesan (2006) performed a very interesting experiment in which the photoconductivity of large area CNT sheets measured upon light illumination. The photoconductivity shows anomalous behavior when light incident on different areas in between the electrodes. Also the effects of light frequency, intensity, and pressure on photoconductivity studied (Fig. 7.5). It was found that photocurrent decreases with an increase in frequency of incident light (Lu and Panchapakesan 2006). Mishra et al. in their article recorded

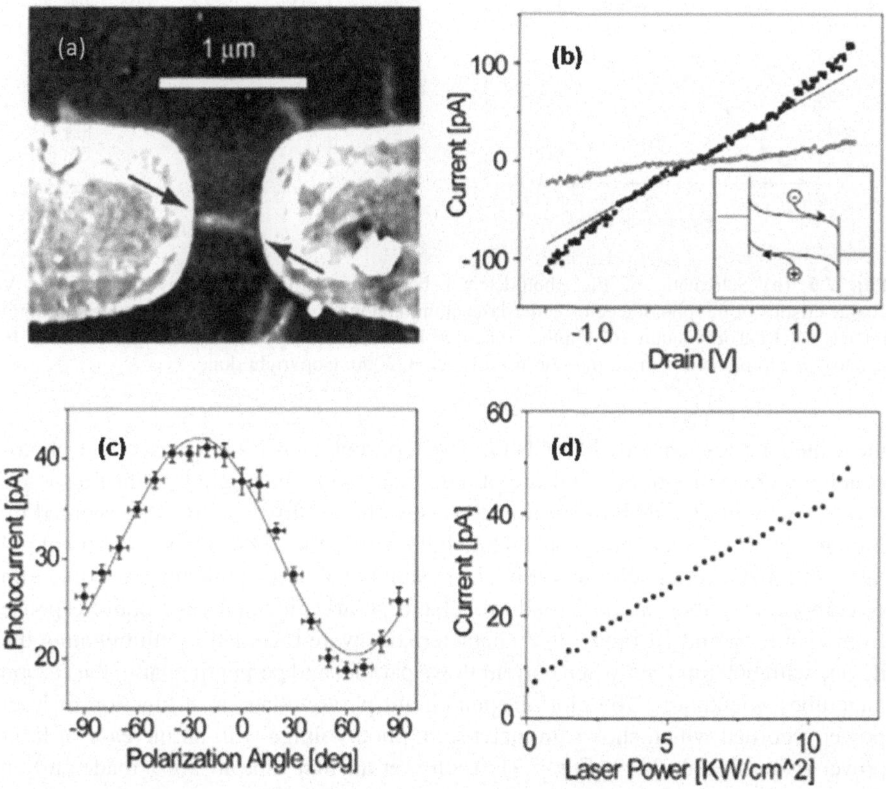

Fig. 7.4 (**a**) SEM image of the single-wall carbon nanotube-based field-effect transistor (**b**) I–V characteristics of the FET with and without laser illumination (**c**) Polarization-sensitive photocurrent of the device under IR laser (**d**) Photocurrent versus laser power plot of the FET. Figure **a**, **b**, **c**, and **d** is reproduced with permission from Freitag et al. 2003. Copyright done

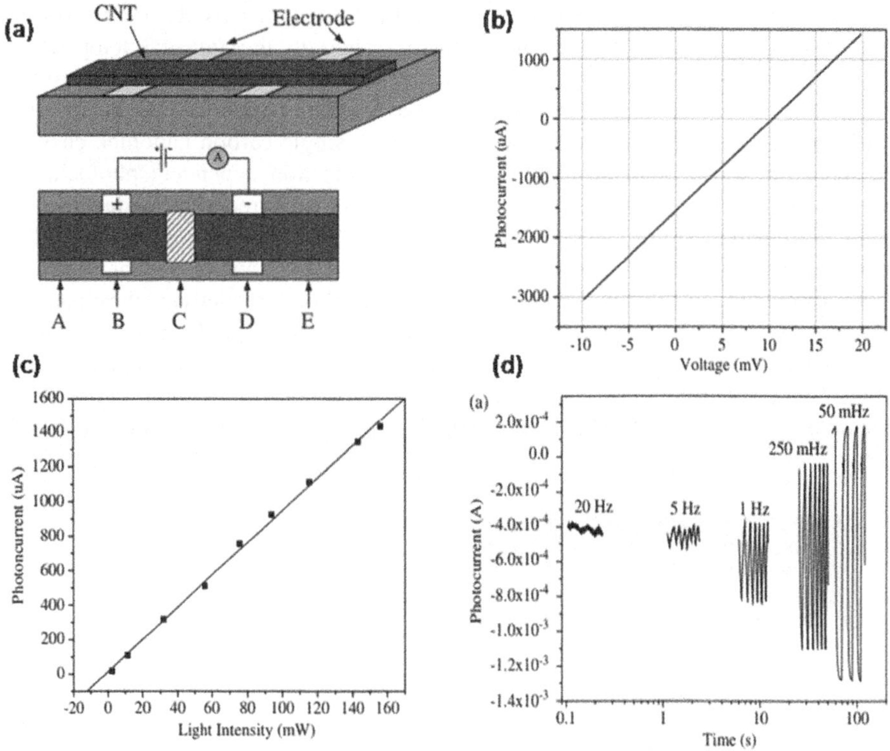

Fig. 7.5 (**a**) Schematic of the photodetector based on carbon nanotubes sheets (**b**) I–V characteristics of the photodetector under laser illumination (**c**) Photocurrent as a function of light intensity at 1 mTorr vacuum (**d**) Photocurrent at a different frequency of incident light. Figure **a**, **b**, **c**, and **d** is reproduced with permission from Lu et al. 2006. Copyright done

the wideband response of MWCNTs. The reported sensor shows excellent photoconductive response upon exposure of red, green, and white light with fast response and recovery time. Graph of sensitivity versus photon flux density was recorded by illuminating it with red, green, and white light, it was found that the sensitivity of red light is highest among all lasers (Fig. 7.6). Nanot et al. aligned the macroscopic film of carbon nanotubes and recorded broadband polarization-sensitive photoresponse from visible to mid-IR range. I–V characteristics were recorded by illuminating the device with 660 nm laser when current flows parallel and perpendicular to the carbon nanotubes orientation. The plot of open circuit photovoltage as a function of laser power recorded which shows an increase in photovoltage with an increase of laser power (Nanot et al. 2013, Fig. 7.7). Tzolov et al. first time in 2007 made carbon nanotubes- silicon heterojunction which responds from near to mid-IR range in both cooled and uncooled modes. The periodic array of carbon nanotubes parallel to each other on the silicon substrate offers a new platform with tenability and scalability. The bandgap of carbon nanotubes determined to form spectral dependence

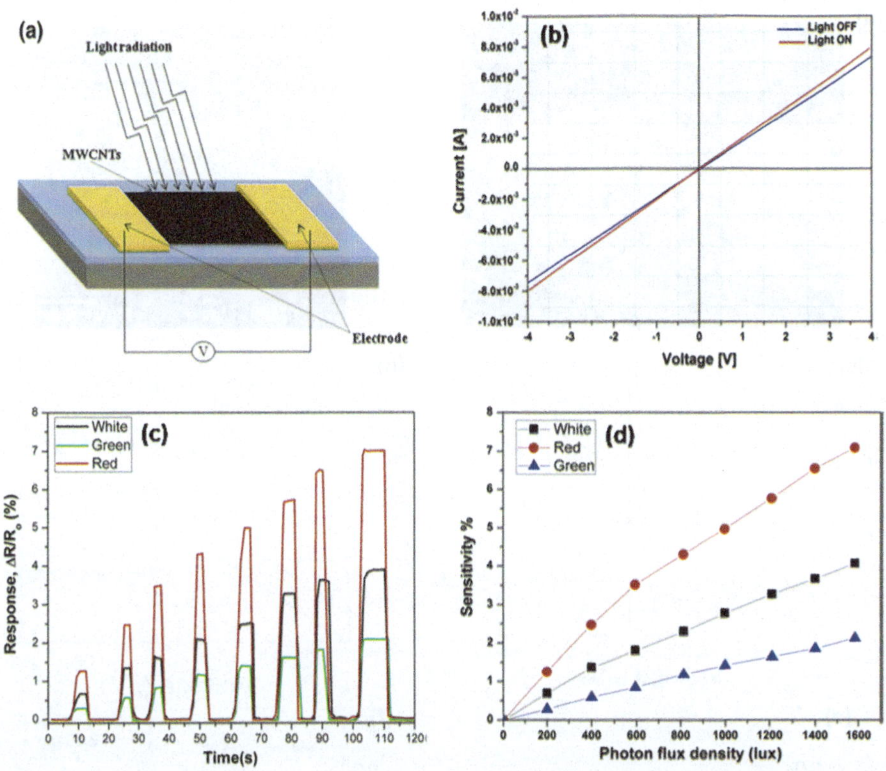

Fig. 7.6 (**a**) Schematic of the photodetector based on MWCNT composite film. (**b**) I–V characteristics of the detector under dark and white light illumination. (**c**) Photoresponse of detector in presence of green, white, and red light. (**d**) Sensitivity versus Photon flux density photoresponse in presence of red, green, and white light. Figure **a**, **b**, **c**, and **d** is reproduced with permission from Mishra et al. 2014. Copyright done

photocurrent and was found in good agreement with the thermally activated measured value. Graph of normalized photocurrent versus excitation energy spectra at 100 K, 180 K, and room temperature were drawn which shows high photocurrent at low temperature (Fig. 7.8) (Tzolov et al. 2007).

Park et al. made a novel transistor by combining single-wall carbon nanotubes and C_{60} with improved photosensitivity and photodetection gate tunability. A device designed in such a way that a large number of generated charge carriers are trapped in it. Different optical parameters like responsivity drawn as a function of the source to drain bias at different gate voltage and graph of photocurrent versus light intensity plotted which shows a linear increase in photocurrent with the light intensity of the phototransistor (Fig. 7.9) (Park et al. 2015).

Baofang Cai et al. made single-wall carbon nanotubes-graphene Schottky junction which shows excellent optoelectronic properties and efficient charge transfer in the device. The device shows broadband photodetection from visible to NIR ranges

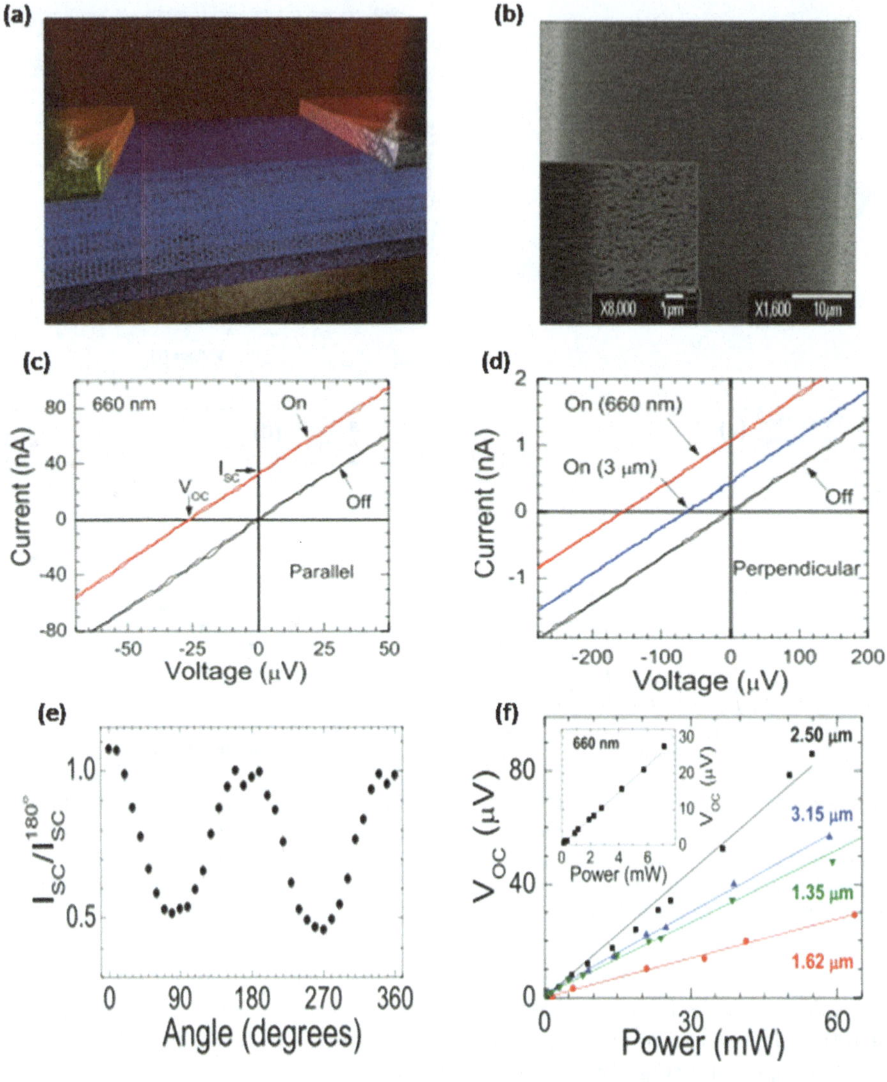

Fig. 7.7 (**a**) Schematic of the photodetector based on aligned carbon nanotubes network (**b**) SEM images of aligned carbon nanotubes (**c, d**) I–V characteristics of the photodetector under laser illumination for the device when current flows parallel and perpendicular to the carbon nanotubes orientation (**e**) Polarization sensitive photoresponse of the detector under 660 nm laser illumination (**f**) Open circuit photovoltage as a function of laser power. Figure **a**, **b**, **c**, **d**, **e,** and **f** is reproduced with permission from Nanot et al. 2013. Copyright done

with very high photoresponsivity and gain achieved so far. Device architecture reduces the recombination capability of charge carriers and opens a new window for carbon nanotubes photodetectors (Fig. 7.10) (Cai et al. 2018). Ma et al. (2019)

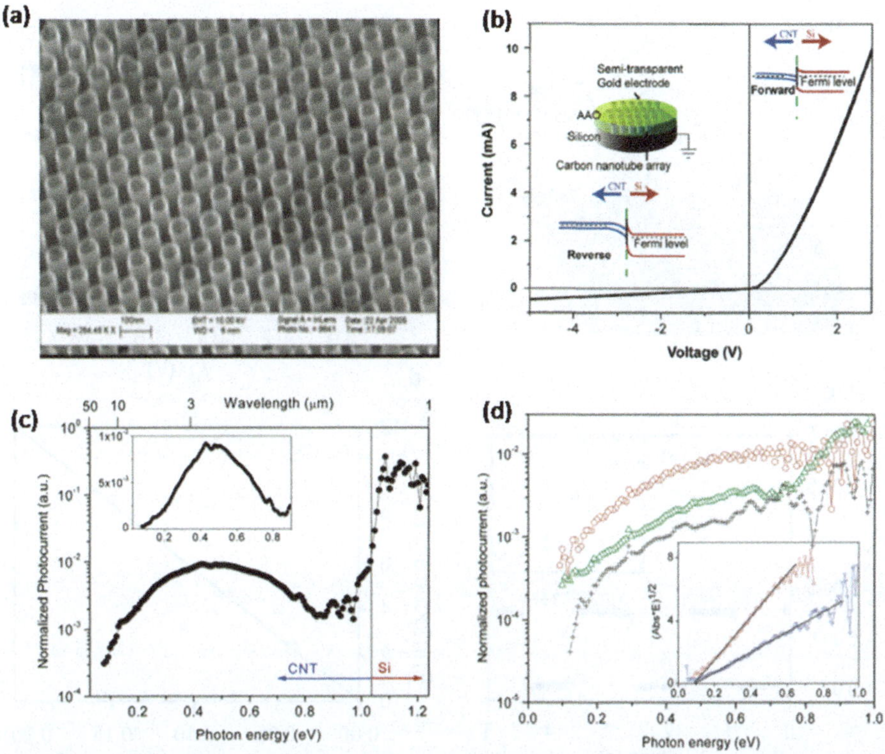

Fig. 7.8 (**a**) SEM image of the carbon nanotube array on the silicon substrate (**b**) I–V characteristics of the Si/carbon nanotubes heterojunction under the dark condition at room temperature (**c**) Normalized photocurrent versus excitation energy spectra at room temperature (**d**) Normalized photocurrent versus excitation energy spectra at 100 K, 180 K, and room temperature. Figure **a**, **b**, **c**, and **d** is reproduced with permission from Tzolov et al. 2007. Copyright done

improved the performance and uniformity of carbon nanotubes based photodiode by coating and decoating of yttrium oxide without introducing any new defects.

The detector shows improved differential resistance at zero bias with a higher signal to noise ratio (Fig. 7.11) (Ma et al. 2019). He et al. reported on the development of a powerless, compact, broadband, flexible, large area, and polarization-sensitive carbon nanotube THz detector that works at room temperature. The I–V characteristics are measured under illumination by linearly polarized THz beam (with frequency of 2.52 THz) in the air at room temperature where current is more than without illumination. The photo signal produced by the detector strongly depends on the polarization of the incident THz beam with respect to the nanotube alignment direction. The power dependence photovoltage shows a linear response and the responsivity of the detector can be determined from the slope of the photovoltage versus power curve (Fig. 7.12) (He et al. 2014).

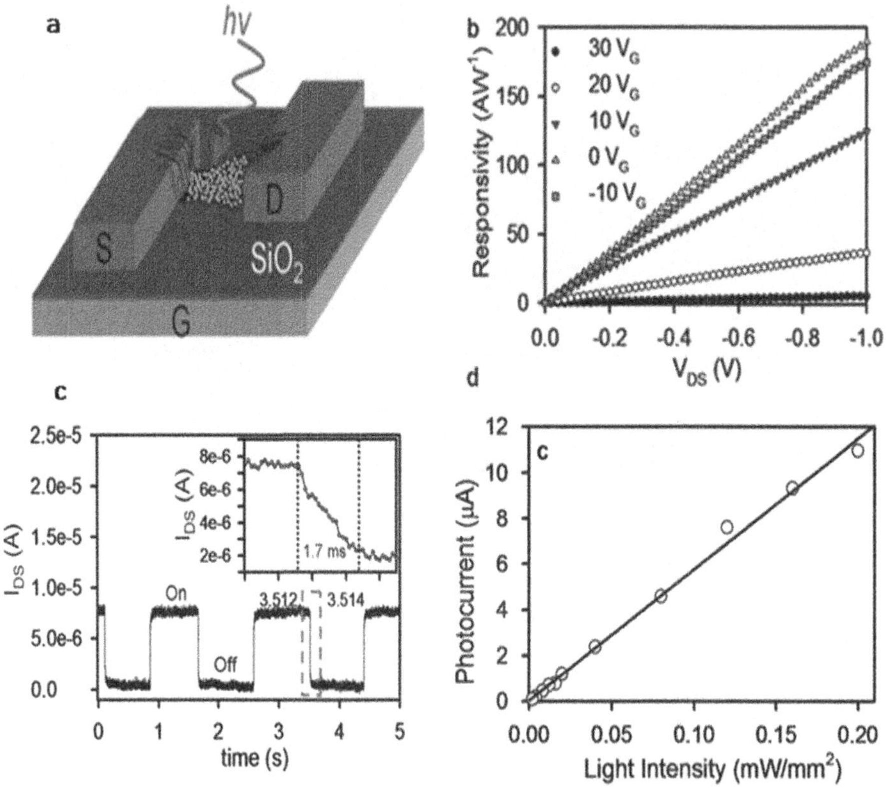

Fig. 7.9 (**a**) Schematic of the phototransistor based on SWNT/C 60 film (**b**) Responsivity as a function of the source to drain bias at different gate voltage (**c**) Current versus time graph at V_{DS} −0.5 V and V_G at 25 V (**d**) Photocurrent versus light intensity plot of the phototransistor. Figure **a**, **b**, **c**, and **d** is reproduced with permission from Park et al. 2014. Copyright done

Liu et al. reported an ultra-broadband photodetectors (Ultraviolet to Terahertz) based on suspended carbon nanotube film. CNT films with an ample distribution of tube diameters and spiderweb-like morphology exhibit a strong absorption spectrum from the ultraviolet up to the terahertz region.

The device showed linear I–V characteristics with 1 V bias voltage range for both in air and a vacuum, as shown in Fig. 7.13a, b. Furthermore, the photocurrent ΔI ($\Delta I = I_{\text{light}} - I_{\text{dark}}$) also display a linear relationship with the same bias voltage. Due to the desorption of air molecules in vacuum dark current is decreased. Figure 7.13c, d demonstrates the photoresponse of the device has been demonstrated in illumination with different wavelengths lasers from the UV to the terahertz regime.

The photocurrent of a device is linearly depend on the power of absorbed light under illumination from Ultraviolet to Terahertz range in the air (in Fig. 7.14) whereas it is not linear in a vacuum (in Fig. 7.14e–h). It can be affirmed that with the same absorbed light power and the absolute value of photocurrent is much larger

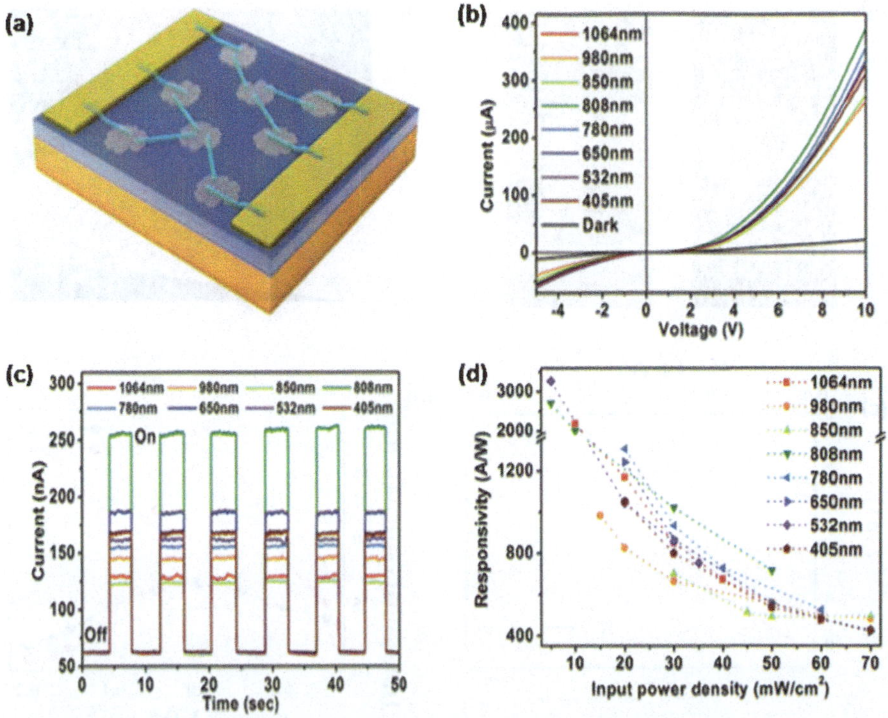

Fig. 7.10 (**a**) Schematic of the photodetector based on SWCNT/SGR hybrid networks (**b**) I–V characteristics of the photodetector at different wavelengths (**c**) Current versus time graph at different wavelengths (**d**) Responsivity versus laser power density at different wavelengths. Figure **a**, **b**, **c**, and **d** is reproduced with permission from Cai et al. 2018. Copyright done

in vacuum than in air. Results determined that CNT films are auspicious for use as the active material in ultra-broadband photodetectors covering the UV to terahertz range of incident light (Liu et al. 2018).

7.2 Conclusion

We have shown advancement in the field of carbon nanotubes based optical detectors. Different types of optical detectors employed using carbon nanotubes working at low temperatures and room temperature. The effects of various parameters such as frequency, pressure, biasing, and doping on optical detection have been studied. The introduction of terahertz technology broadens the detection band of carbon nanotubes from UV to THz of the electromagnetic spectrum. Several device structures, heterojunctions, and composites were proposed for improvements in the photoresponse, stability, repeatability, and uniformity of detectors. Optical

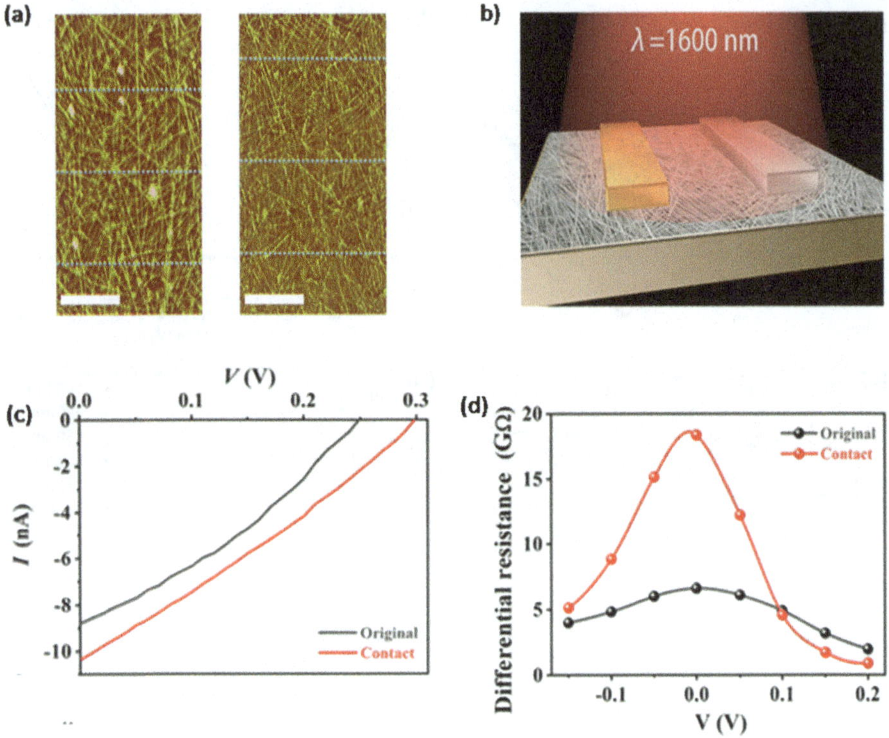

Fig. 7.11 (**a**) AFM images of the CNT thin film before and after YOCD treatment (**b**) Schematic of the CNT photodiode (**c**) I–V characteristics of the detector with and without contact area treatment (**d**) Differential resistance versus bias voltage in the presence of 1600 nm laser. Figure **a**, **b**, **c**, and **d** is reproduced with permission from Ma et al. 2019. Copyright done

response in the UV range is due to the π-plasmon, from visible to NIR photoresponse is due to the interband transitions, and from mid-IR to THz photoresponse is caused by the plasmonic resonance in the carbon nanotubes. Modification in optical properties and device engineering in carbon nanotubes based detector will lead to the development of novel photodetectors.

The properties of CNTs are quite encouraging in the detection of THz range. Detection of THz range by CNTs already demonstrated by few research groups still they did not propose proper physics and technology but they developed a path towards THz detection. Since the potential and path toward improvement have been clearly established, so, researchers have lots of scope on this field.

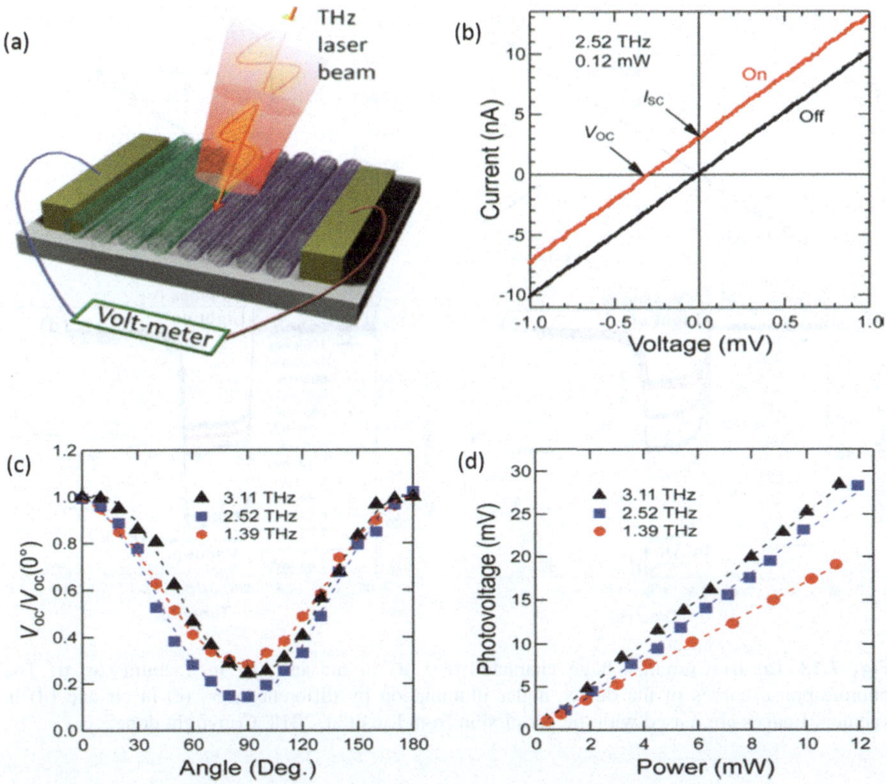

Fig. 7.12 (**a**) A schematic diagram of the device. (**b**) I–V characteristics under illumination (red) and without illumination (black). (**c**) Polarization-sensitive photovoltage (**d**) Power dependence photovoltage. Images are reused with the permission of He et al. 2014. Copyright done

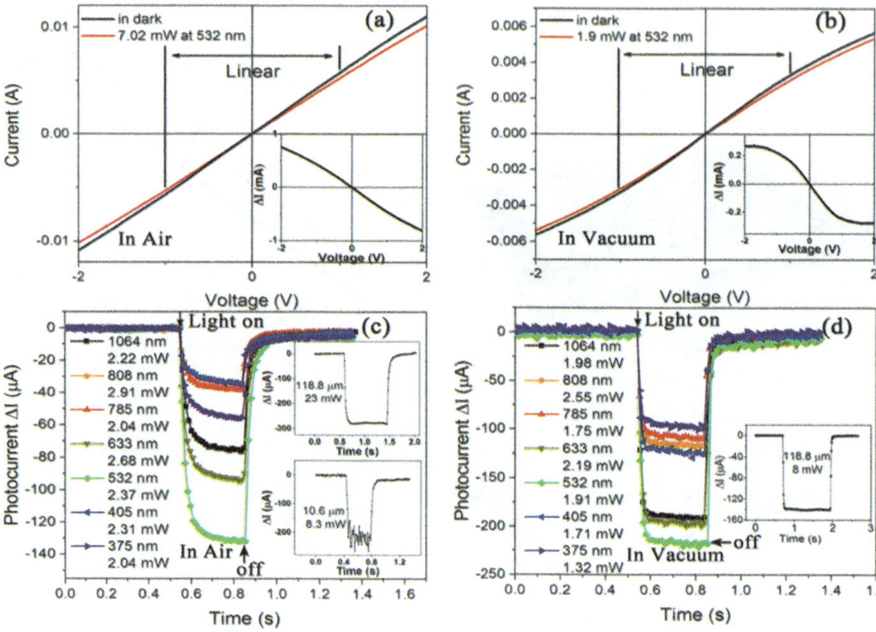

Fig. 7.13 (**a**, **b**) Current-voltage characteristics (**a**) in air and (**b**) in vacuum. (**c**, **d**) The photoresponse curves of the device under illumination by different lasers (**c**) in air and (**d**) in vacuum. Images are reused with the permission from Liu et al. 2018. Copyright done

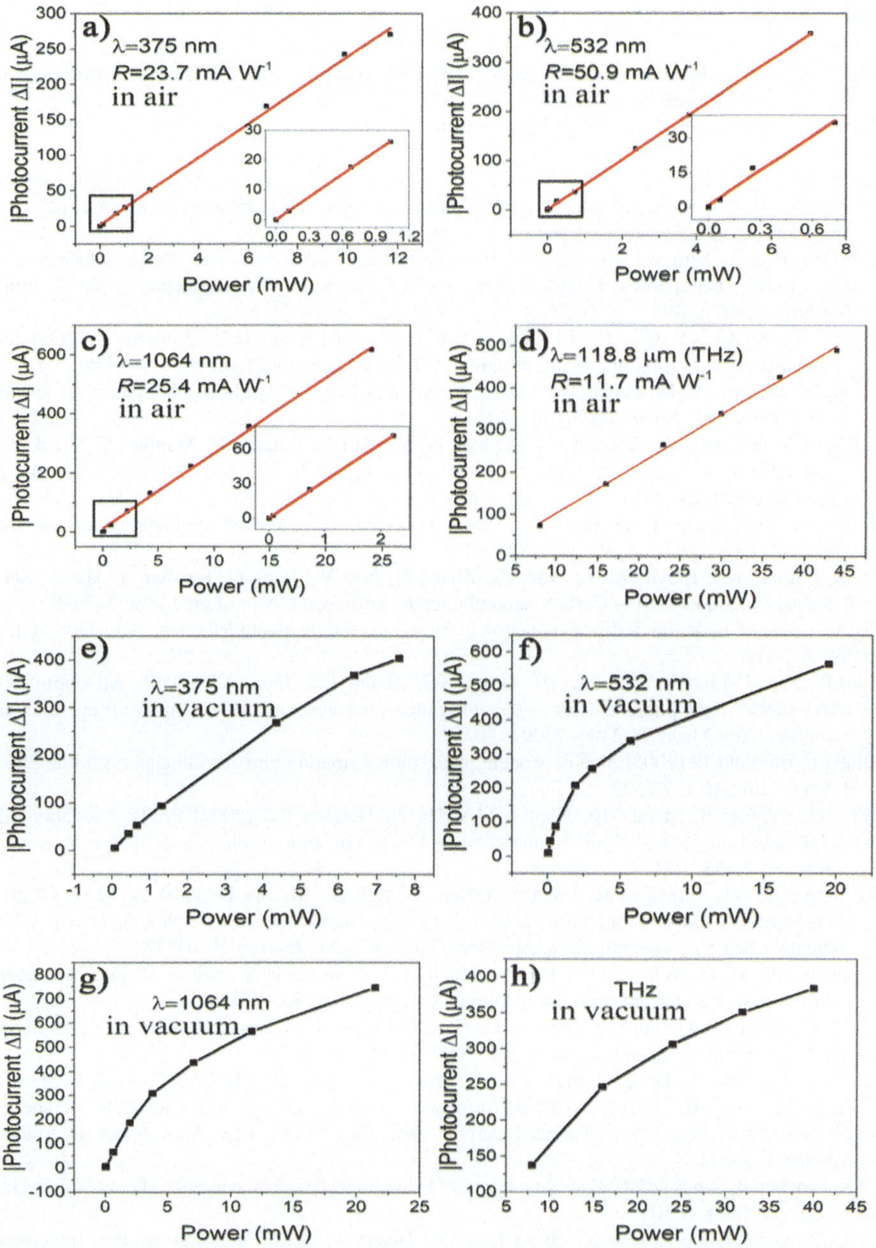

Fig. 7.14 Photocurrent |Δ*I*| versus the absorbed light power (**a–d**) in air and (**e–h**) in vacuum. Images are reused with the permission from Liu et al. 2018. Copyright done

References

Avouris P, Freitag M, Perebeinos V (2008) Carbon-nanotube photonics and optoelectronics. Nat Photonics 2:341–350

Bonaccorso F, Sun Z, Hasan T (2010) Graphene photonics and optoelectronics. Nat Photonics 4:611–622

Cai B, Su Y, Tao Z, Hu J, Zou C, Yang Z, Zhang Y (2018) Highly sensitive broadband single-walled carbon nanotube photodetectors enhanced by separated graphene nanosheets. Adv Opt Mater 6:1800791

Choi MS, Park T, Kim WJ, Hur J (2020) High-performance ultraviolet photodetector based on a zinc oxide nanoparticle@single-walled carbon nanotube heterojunction hybrid film. Nanomaterials 10:395

Dai JC, Xu X, Xu XY, Guo JY, Li JT, Zhu GY, Lin Y (2013) Single ZnO microrod ultraviolet photodetector with high photocurrent gain. ACS Appl Mater Interfaces 5:9344–9348

Freitag M, Martin Y, Misewich JA, Martel R, Avouris PH (2003) Photoconductivity of single carbon nanotubes. Nano Lett 3:1067–1071

Fujiwara A, Matsuoka Y, Suematsu H, Ogawa N, Miyano K, Kataura H, Maniwa Y, Suzuki S, Achiba Y (2001) Photoconductivity in semiconducting single-walled carbon nanotubes. Jpn J Appl Phys 40:1229–1231

Hartmann RR, Kono J, Portnoi M (2014) Terahertz science and technology of carbon nanomaterials. Nanotechnology 25:322001

He X, Fujimura N, Lloyd JM, Erickson K, Zhang Q, Gao W, Jiang Q, Kawano Y, Hauge RH, Leonard F, Kono J (2014) Carbon nanotube terahertz detector. Nano Lett 14:3953–3958

He X, Leonard F, Kono J (2015) Uncooled carbon nanotube photodetectors. Adv Opt Mater 3:989–1011

Hu LF, Yan J, Liao MY, Xiang HJ, Gong XG, Zhang LD, Fang XS (2012) An optimized ultraviolet-A light photodetector with wide-range photoresponse based on ZnS/ZnO Biaxial Nanobelt. Adv Mater 24:2305–2309

Huang F, Banfield JF (2005) Size-dependent phase transformation kinetics in nanocrystalline ZnS. J Am Chem Soc 127:4523

Itkis ME, Niyogi S, Meng ME, Hamon MA, Hu H, Haddon RC (2002) Purity evaluation of As-prepared single-walled carbon nanotube soot by use of solution-phase near-IR spectroscopy. Nano Lett 2:155–159

Ka I, Borgne VL, Fujisawa K, Hayashi T, Kim YA, Endo M, Ma D, Khakani MAE (2020) PbS-quantum- dots/double-wall-carbon-nanotubes nanohybrid based photodetectors with extremely fast response and high responsivity. Mater Today Energy 16:100378

Levitsky IA, Euler WB (2003) Photoconductivity of single-wall carbon nanotubes under continuous-wave near-infrared illumination. Appl Phys Lett 83:1857

Li Z, Salfi J, Souza CD, Sun P, Nair SV, Ruda HE (2010) Van der Waals epitaxial ZnTe thin film on single-crystalline graphene. Appl Phys Lett 97:063510

Liu Y, Yin J, Wang P, Hu Q, Wang Y, Xie Y, Zhao Z, Dong Z, Zhu JL, Chu W, Yang N, Wei J, Ma W, Sun JL (2018) High-performance, ultra-broadband, ultraviolet to terahertz photodetectors based on suspended carbon nanotubes films. ACS Appl Mater Interfaces 10:36304–36311

Lu S, Panchapakesan S (2006) Photoconductivity in single wall carbon nanotube sheets. Nanotechnology 17:1843–1850

Lu Q, Zeng H, Wang Z, Cao X, Zhang L (2006) Design of Sb2S3 nanorod-bundles: imperfect oriented attachment. Nanotechnology 17:2098–2104

Ma Z, Han J, Yao S, Wang S, Peng LM (2019) Improving the performance and uniformity of carbon-nanotube-network-based photodiodes via yttrium oxide coating and decoating. ACS Appl Mater Interfaces 11:11736–11742

Majumder R, Kundu S, Ghosh R, Pradhan M, Ghosh D, Roy S, Roy S, Chowdhury MP (2020) Expeditious UV detection of tungstite ($WO_3{\cdot}H_2O$) and tungsten oxide (WO_3) decorated multiwall carbon nanotubes (MWCNT) based photodetector: ultrafast response and recovery time. Appl Sci 2:81

Mishra P, Harsh, Islam SS (2014) Development of MWCNTs-based wideband photodetector in the visible range: wavelength and power-dependent response studies. Appl Phys A 117:1119–1123

Nanot S, Cummings AW, Pint CL, Ikeuchi A, Akiho T, Sueoka K, Hauge RH, Leonard F, Kono J (2013) Broadband, polarization-sensitive photodetector based on optically-thick films of macroscopically long, dense, and aligned carbon nanotubes. Sci Rep 3:1335

Ostojic G, Zaric S, Kono J, Strano M, Moore V, Hauge R, Smalley R (2004) Interband recombination dynamics in resonantly excited single-walled carbon nanotubes. Phys Rev Lett 92:117402

Park S, Lee SR, Tung L et al (2014) Optical mapping of optogenetically shaped cardiac action potentials. Sci Rep 4:6125

Park S, Kim SJ, Nam JH, Pitner G, Lee TH, Ayzner AL, Wang H, Fong SW, Vosgueritchian M, Park YZ, Brongersma ML, Bao Z (2015) Significant enhancement of infrared photodetector sensitivity using a semiconducting single-walled carbon nanotube/C_{60} phototransistor. Adv Mater 27:759–765

Pichler T, Knupfer M, Golden MS, Fink J, Rinzler A (1998) Localized and delocalized electronic states in single-wall carbon nanotubes. Phys Rev Lett 80:4729–4732

Posada FG, Songmuang R, Hertog MD, Monroy E (2012) Room-temperature photodetection dynamics of single GaN nanowires. Nano Lett 12:172–176

Ren L, Pint CL, Arikawa T, Takeya K, Kawayama I, Tonouchi M, Hauge RH, Kono J (2012) Carbon nanotube terahertz polarizer. Nano Lett 12:787–790

Spataru CD, Leonard F (2010) Tunable band gaps and excitons in doped semiconducting carbon nanotubes made possible by acoustic plasmons. Phys Rev Lett 104:177402

Svensson J, Anttu N, Vainorius N, Borg BM, Wernersson LE (2013) Diameter-dependent photocurrent in InAsSb nanowire infrared photodetectors. Nano Letter 13:1380–1385

Talib M, Tabassum R, Islam SS, Mishra P (2019) Influence of growth temperature on titanium sulphide nanostructures: from trisulphide nanosheets and nanoribbons to disulphide nanodiscs. RSC Adv 9:645–657

Tzolov MB, Kuo TF, Straus DA, Yin A, Xu J (2007) Carbon nanotubes-silicon heterojunction arrays and infrared photocurrent responses. J Phys Chem C 111:5800–5804

Wang ZX, Safdar M, Jiang C, He J (2012) ZnO nanowire UV photodetectors with high internal gain. Nano Lett 12:4715–4721

Wu PC, Dai Y, Sun T, Ye Y, Meng H, Fang XL, Yu B, Dai L (2011) Monolayer graphene/germanium Schottky junction as high-performance self-driven infrared light photodetector. ACS Appl Mater Interfaces 3:1859–1864

Yan RX, Gargas D, Yang PD (2009) Nanowire photonics. Nat Photonics 3:569–576

Zhanga Y, Denga T, Lia S, Suna J, Yina W, Fanga Y, Liu Z (2020) Highly sensitive ultraviolet photodetectors based on single wall carbon nanotube-graphene hybrid films. Appl Surf Sci 512:145651

Zheng W, Huang F, Zheng R, Wu H (2015) Low-dimensional structure vacuum-ultraviolet-sensitive ($\lambda < 200$ nm) photodetector with fast-response speed based on high-quality AlN micro/nanowire. Adv Mater 27:3921

Zhou W, Peng Y, Yin Y, Zhou Y, Zhang Y, Tang D (2014) Channel length specific broad spectral photosensitivity of robust chemically grown CdS photodetector. AIP Adv 4:123005

Zhou H, Wang J, Ji C, Liu X, Han J, Yang M, Gou J, Xu J, Jiang Y (2019) Polarimetric Vis-NIR photo detector based on self-aligned single-walled carbon nanotubes. Carbon 143:844–850

Recent Advances of Fluorescence Resonance Energy Transfer-Based Nanosensors for the Detection of Human Ailments

8

Amreen, Mohammad Ahmad, and Ruphi Naz

Abstract

This chapter enlightens the development and application of dye-based as well as genetically encoded nanosensors for metabolites related to various diseases, their comparison, and usefulness in understanding the metabolic pathways by using novel high throughput assays. Previously dye-based fluorescent probes have been constructed for the in vivo detection of cellular molecules. To study the real-time localization and monitoring of metabolites in live cells, genetically encoded fluorescent nanosensors are steadily used as an elected method in comparison with chemical probes that are intrinsically dependent on effective delivery into the cell and sometimes causes the toxicity in the cells. Several methods have been reported to detect cancer and neurodegenerative diseases. However, FRET-based disease-specific biomarker detection is a preferable method because of its high sensitivity and selectivity. By using FRET-based genetically encoded nanosensor, we can detect some of the metabolites which are related to several diseases like cancer, Alzheimer, and diabetes. In humans, lack of reactive oxygen species (ROS) causes certain autoimmune diseases and excess production of ROS causes cardiovascular and respiratory diseases which can be detected in real time and non-invasively by using a nanosensor. Protein recognition-based genetically encoded sensors are projected to discover many research, medical, and environmental applications. Many human problems can be solved by the utilization of nanosensors which can also be helpful to authenticate the medicines to their

Amreen (✉)
Department of Botany, Aligarh Muslim University, Aligarh, UP, India

M. Ahmad
Department of Botany, Hamdard University, New Delhi, India

R. Naz
Department of Biosciences, Jamia Millia Islamia, New Delhi, Delhi, India

M. Mohsin et al. (eds.), *Nanobiosensors for Agricultural, Medical and Environmental Applications*, https://doi.org/10.1007/978-981-15-8346-9_8

principle components. Thus, sensor biology can be revolutionary for human welfare in the future with the help of FRET-based tools.

Keywords

Nanosensor · Diseases · FRET · ROS

8.1 Introduction

After the sequencing of the genome of model organisms, it is now important to assign the function to every gene present in the genome, and their incorporation into metabolic and regulatory pathways. While transcriptomics and proteomics are developing rapidly, the collection of the necessary information for constructing these network maps is being developed. Metabolomic technologies supplement these methods by testing for fluctuations in cellular ion and metabolite patterns, providing necessary information for the building of cellular metabolic networks. Despite the increasing progress in the fields of transcriptomics and proteomics, understanding of the regulatory networks controlling the metabolic pathway is poorly known. Major components of the transport machinery are still unknown. One of the major problems in accurately defining the metabolic network is the fact that metabolites are highly compartmentalized into cellular and subcellular organelles. Biochemical pathways are commonly distributed between compartments and even adjoining cells or tissues. Thus, metabolite levels in a group of neighboring cells and tissues are not essentially the same, as mentioned for glucose and lactate in the liver (Arlt et al. 2001). Because metabolism is synchronized at many levels, the investigation of all regulatory levels is a pre-requisite for broad network analysis. Metabolic flux is the movement of metabolite within the metabolic pathway. Each flux indicates the function of a particular pathway within the network. It is the flux that delivers the phenotype of an organism. An interesting significance of the regulatory structure of pathways is that the major fluctuations in flux are sometimes linked with only uncertain adjustments in metabolite levels. This indicates that metabolomics estimation of metabolite composition may be unsatisfactory for a complete knowledge of the metabolic phenotype; therefore, flux measurement could offer a valuable complementary parameter for the characterization of metabolic networks. The need to understand the metabolic network is likely to be increased importance for metabolic engineering. Keeping aside the most minor fluxes, such as uptake and secretion rates, yet it is not possible to monitor fluxes directly in a metabolic network; they can be inferred from stoichiometric models. Biological systems comprise an immense number of specific components, which undergo lively and highly communicating responses in time and space. Exploration of these responses will offer the key to release the information encoded in the genome sequences.

With the currently available techniques, we can only get partial pictures of the highly dynamic alterations of metabolite levels with poor spatial resolution. For metabolomics studies, there is a need of the tools for (1) the determination of the level of a specific molecular species in a physiological environment, (2) the detection of the changes in metabolite level in response to environmental stimuli, (3) the measurement of multiple analytes and (4) the measurements of a variety of structural and functional analytes. Several groups have started to develop the genetically encoded FRET-based sensors by using a combination of fluorescent proteins (FPs). Such kind of nanosensors have been developed for amino acids (Okimono et al. 2005; Bogner and Ludewig 2007; Mohsin et al. 2013), sugars (Fehr et al. 2003; Lager et al. 2006), and many other metabolites (Kaper et al. 2007; Ellis and Wolfgang 2012; Song et al. 2015). The genetically encoded nanosensor can work non-invasively and monitor the specific analyte in real-time. Genetically encoded nanosensor based on FRET is a promising technique to fulfill this demand. The FRET-based sensors have been designed by exploiting the GFP or its variants, which are fused to ligand-specific domains in a manner that makes the fluorescent signal sensitive to specific changes in the local environment. Different nanosensors have so far been created by using the circularly permutated fluorescent protein (cpFP), in which the N and C termini were changed by a short peptide linker that altered the pKa values and orientation of chromophore from their novel counterpart (Baird et al. 2000). This strategy (cpFP) has been used in the designing of calcium sensor, pericam (Nagai et al. 2001) and ATP indicator, ATeams (Imamura et al. 2009). Genetically encoded FRET-based sensor is a tandemly combined protein that detects the target metabolite based on the conformational alteration in the sensory domain. Changes in the conformation of the domain result in the form of FRET between two fluorescent moieties. Attachment of the ligand to the sensory domain will increase energy transfer efficiency from donor FP to the acceptor FP. The intensity ratio of two FPs reports the fluctuations in the concentration of target metabolites. The FRET is highly effective if the donor and acceptor fluorophores are situated within the radius and can be a measurement of molecular closeness at nanometer distances. FRET efficiency depends on some physical conditions. It is based on the inverse sixth power of intermolecular separation (Lakowicz 1999), spectral overlapping, and orientation of the fluorophores. It is a complex physical method for exploring various biological events that produce alterations in molecular proximity (Dos Remedios et al. 1987). In previous years, FRET-based sensor approach was effectively used for the in vivo detection of calcium, glucose, matrix metalloproteinase (MMP-9) activity, and abscisic acid dynamics (Miyawaki et al. 1997; Fehr et al. 2003; Stawarski et al. 2014; Jones et al. 2014). Fluorescence resonance energy transfer-based biosensor has also been developed to analyze the activity of focal adhesion kinase that regulates cell migration, adhesion, and proliferation (Seong et al. 2011). This chapter explains the advancement and applications of genetically encoded as well as dye-based nanosensors for metabolites related to various diseases and their usefulness in understanding the metabolic and regulatory pathways.

8.2 Nanosensor as a Sensing Tool

Nanosensor is a device which uses biological recognition elements to transform a cellular process into detectable form to do the measurement on a sample and can be analyzed in real-time condition. This device must also then change that measurement into a reportable signal which may be an electrical signal, chemical signal, luminescent signal, or fluorescent signal. Different recognition of biological molecules as recognition elements have been used for the development of different variety of biosensors. Based on the nature of the recognition element or signal transduction process biosensors can be distinguished. As a recognition element DNA, RNA, and protein can also be used to measure the concentration of metabolites within its host or extracellular environment (Song et al. 2012). For the quantification of target molecule concentration, protein-based biosensor is an interesting class of biosensor that utilizes FRET as a signal transduction process.

The main components of FRET-based sensor are (1) Fluorescent protein, (2) FRET phenomenon, and (3) Ligand-binding protein.

8.2.1 Fluorescent Proteins (FPs)

For the designing of the FRET-based sensor, three major types of fluorophores have been exploited as FRET pairs: fluorescent proteins, quantum dots, and small organic dyes. Not like quantum dots and dyes, FPs are genetically encoded, which can be advantageous in live cell FRET imaging. The best known fluorescent protein is a Green fluorescent protein (GFP) which was isolated from Aequorea Victoria (Tsien 1998). In current cell biology research, FPs are now the fundamental tool because of their several advantages (1) Fusing FPs simply to sensing domains through genetic engineering, the FPs based FRET sensors can be easily constructed. In contrast, dyes and QDs need antibodies to label sensing domains. (2) Through transfection or virus infection FPs based sensors can be easily introduced into the cells, while to introduce quantum dots and dye-based FRET sensor inside the cell is challenging (3) Because of the high intracellular stability of FPs, nanosensors based on FPs are stable in cells for a long time but dye- and QD-based sensors are unstable in living cells (Lesmana and Friedl 2001).

8.2.2 Forster Resonance Energy Transfer (FRET)

In 1948, the concept of FRET was originated by Theodor Forster. It is a distance-dependent phenomenon describing energy transfer between two fluorophores of different colors. In which, a donor fluorophore, in its excited state transfer energy to an acceptor fluorophore non-radiatively through dipole–dipole coupling (Forster 1965). It is extremely sensitive to the small changes in distance between the donor and the acceptor molecules. Because of this property, FRET has been proved advantageous to study intermolecular interactions.

There are many conditions for FRET to occur:

1. The primary condition is the distance between the donor and acceptor fluorophore should not be ≥ 10 nm.

2. The Excitation spectrum of the acceptor should overlap the emission spectrum of the donor fluorophore.
3. The two fluorophores dipole orientation must be parallel if the orientation is different then no FRET will occur.

8.2.3 Ligand-Binding Protein

The development of FRET-based nanosensor requires the selection of suitable substrate/ligand-binding protein. Substrate binding proteins form a group of proteins that are normally linked with membrane protein complexes designed for the transport of substrate or signal transduction. Substrate binding protein was initially found to be linked with prokaryotic ABC-transporters (Berger 1973; Berger and Heppel 1974; Tam and Saier 1993). L-arabinose was the first ligand-binding protein crystal structure elucidated in 1974 (Quiocho et al. 1974). Since then many other ligand-binding proteins have been elucidated. The ligand-binding protein consists of two α/β domains which are linked by hinge-region, through the ligand-binding site existing in between these two domains. In the absence of ligand, both the domains separated and exist largely in the open conformation (Quiocho and Ledvina 1996). Upon binding of the ligand, both domains come close to each other and the protein shows the closed conformation through trapping the ligand (Mao et al. 1982).

8.3 Noninvasive Tool for Detection of Various Metabolites Which Are Related to Human Health

8.3.1 FRET-Based Sensor for the Detection of Cancer

For the majority of deaths, cancer is the main cause of mortality all around the globe and the second major cause of death after stroke and coronary heart diseases (Fitzmaurice et al. 2018). In recent times, cancer is increasing rapidly due to unhealthy lifestyle choices like smoking, westernized diets, and physical inactivity. About 18.1 million recent cancer cases and 9.6 million cancer deaths are estimated to have occurred in 2018 worldwide (Bray et al. 2018). Researchers have been working to improve reliable systems for the initial detection of cancer. There are various methods for the detection of cancer like ultrasound, X-rays, positron emission tomography, and computerized tomography. However, FRET-based sensors are powerful tools that can detect various enzymes and metabolites related to cancer and other diseases.

8.3.1.1 Dyes Based Sensor for the Detection of Cancer

Cationic Conjugated Polymer (CCP) Based FRET Technique

In somatic tissue, changes in DNA sequences due to methylation and covalent modifications of histone causes cancer. These DNA alterations exert a major effect on gene expression which may remove the influence of tumor suppressor gene products or increase the function of proto-oncogenes (Herman and Baylin 2003; Payne and Kemp 2005). Abnormalities of genomic DNA causes methylation which includes hypomethylation and hypermethylation of promoter of cancer-related genes. Carcinogenesis results from the effect of mutations of TSGs or proto-oncogenes which occurs due to the changes in the methylation of the promoter. Transcriptional silencing of tumor suppressor genes occurs through hypermethylation of the promoter or first exon of the gene related to cancer. Although, the activation of transcription of retrotransposons and proto-oncogenes occurs through the hypomethylation of regulatory DNA sequences. Zhang et al. (2013) developed a Cationic Conjugated polymer (CCP) based FRET technique by which the detection of cancer inside the human cells can occur through analysis of DNA methylation. Methylation sensitive restriction endonucleases were used for the digestion of genomic DNA and then fluorescein-labeled dNTP incorporated through PCR amplification. For the screening and diagnostic of cancer, CCP-based FRET technique has been a powerful tool which can detect methylation of gene promoter at low levels through FRET between CCP and fluorescein which was integrated into the PCR product. Ovarian cancer and normal samples were studied through the detection of methylation levels of their promoters. This detection method takes about 20 h and illustrates vast potential as a beneficial tool for the analysis and testing of cancer (Zhang et al. 2013).

8.3.1.2 Genetically Encoded FP-Based Sensors for the Detection of Cancer

FRET-Based Sensor for BCR-ABL Kinase Activity

Chronic myeloid leukemia, is a cancer of white blood cells marked by excessive proliferation of granulocytic cell lines. It is a form of leukemia categorized by the presence of the Philadelphia chromosome (Kurzrock et al. 1988). BCR-ABL, a novel fusion gene formed from the reciprocal translocation between the chromosomes 9 and 22, which results in the growth of the Philadelphia chromosome. BCR-ABL tyrosine kinase is responsible for the pathogenesis of chronic myeloid leukemia (Groffen et al. 1984; Heisterkamp et al. 1982). FRET-based nanosensor has been developed for the measurement of BCR-ABL activity inside the cells by using CrkL, a substrate of BCR-ABL, which is involved in various pathophysiological processes such as transformation, proliferation of cells, migration, and differentiation (Feller 2001). CrkL consists of one SH2 and two SH3 domains along with a tyrosine residue which is phosphorylated by the tyrosine kinases (Mizutani et al. 2010). This sensor was constructed by sandwiching the CrkL between the two fluorescent proteins namely Venus and eCFP to the C- and N-terminal ends of CrkL. This chimeric protein was designated as Pickles (phosphorylation indicator of CrkL en substrate). In the presence of BCR-ABL, the tyrosine residue phosphorylated and binding of phosphorylated tyrosine to the SH2 domain brings an increase in the FRET efficiency from eCFP to Venus and can be detected as a change in FRET ratio.

Lactate Sensor

In prokaryotic and eukaryotic cells, lactate plays an essential role in cellular metabolism and have a pathological role in mammalian cells (San Martin et al. 2013). Pathological roles include tumor, progressive loss of structure or function of neurons, and inflammation (Ganapathy et al. 2009; Vander Heiden et al. 2009; Tennant et al. 2010; Funfschilling et al. 2012; Lee et al. 2012). In the absence or low concentration of oxygen, tumor cells show fermentation of glucose which produces a large amount of lactose, this process is known as the Warburg effect. Lactate has an essential role in cellular pathways and is essential in brain tissue activity for the formation of long-period memory (Suzuki et al. 2011; Newman et al. 2011) and neurovascular signaling (Shimizu et al. 2007). FRET-based genetically encoded biosensor laconic has been developed to detect the level of lactate inside the tumor cells which is based on bacterial transcriptional regulator LIdR (San Martin et al. 2013). For the generation of this sensor the genes were selected from *Corynebacterium glutamicum* and *Escherichia coli*, and the fluorescent proteins mTFP and Venus were used as a FRET pair. This sensor was developed by sandwiching the LIdR transcriptional regulator between the two fluorescent proteins mTFP and Venus located at the N and C termini respectively which can detect the concentration of lactate from 1 μM to 10 μM inside the mammalian cells with high specificity and not affected by related metabolites like pyruvic acid, glutamic acid, malic acid, and oxaloacetic acid. Laconic sensor along with the conjugation of monocarboxylate transporter was also able to differentiate that the cell is producing the lactate or consuming the lactate. By using the combination of Monocarboxylate transporter it was also detected that there was three- to fivefold increase in the production of lactate in glioblastoma cells as compared to normal astrocytes and the accumulation of lactate inside the glioma cells was also lower than that of the astrocytes.

8.3.2 FRET-Based Sensor for Neurotransmitters

Neurotransmitters are class of molecules that are utilized by cells to exchange cellular information (Polo and Kruss 2016). The most important role of neurotransmitters is biochemical neurotransmission in the brain. Hence, to know how the brain works and also to treat diseases, neurotransmitters analytical chemistry provides various insights. Functional magnetic resonance imaging and Positron emission tomography are noninvasive tools and can detect molecules in vivo. However, they lack spatiotemporal resolution which is essential to study neurotransmitters around single cells (Meyer-Lindenberg 2010; Lee et al. 2014). Various electrochemical and optical techniques have been developed for the recognition of several neurotransmitters (glutamate, γ-aminobutyric acid (GABA), glycine, norepinephrine) in which FRET-based nanosensor are the promising tools which can detect neurotransmitters non-invasively with high spatiotemporal resolution. Genetically encoded fluorescent sensors for glutamate have also been developed by using the bacterial glutamate binding protein along with two fluorescent proteins YFP and CFP. When ligand (glutamate) binds to the glutamate binding protein, the conformation of the protein changes and change in FRET occurs. Hence, to detect glutamate concentration inside the cells change of the FRET signal can be used. This sensor was named as fluorescent indicator protein for glutamate (FLIPE) (Okumoto et al. 2005). Masharina et al. developed the FRET-based GABA sensor which senses GABA with high specificity on the surface of living mammalian cells. This sensor consists of a fusion protein between SNAP-tag, CLIP-tag, and a receptor protein (RP) for the analyte of interest. In living cells, the ratiometric change of this sensor is 1.8 and it can sense micromolar to millimolar concentration of GABA (Masharina et al. 2012). An optical sensor was also developed to detect glycine concentration in the brain. This sensor was developed by using the *Agrobacterium tumefaciens* protein Atu2422 which binds to glycine (Glee), l-serine (Ser), and GABA. Aiming to form binding specificity, this protein was mutated at the sites which are responsible for binding to serine and GABA. This sensor was developed by cloning the AYW molecule between the FRET pair Venus and eCFP. When glycine binds to this protein the conformational changes occur which alters the fluorescent FRET signal (Rusakov 2018). Cell-based neurotransmitter fluorescent engineered reporters (CNiFERs) were also reported to detect dopamine (DA) and norepinephrine (NE). They selected two GPCRs for the detection of NE and DA with high affinity and selectivity: the α_{1A} adrenergic receptor and the D2 dopaminergic receptor that couple to endogenous Gq proteins. Activation of GPCRs by the phospholipase C-inositol triphosphate (PLC-IP3)pathway triggers a rise in cytosolic [Ca^{2+}] level, lead to an increase in FRET which is the rapid optical information of the change in neurotransmitter levels (Muller et al. 2014).

8.3.3 FRET-Based Sensor for Diabetes

8.3.3.1 Glucose Sensor

Diabetes, a devastating metabolic disorder which results from high blood glucose level when it rises more than normal (Piero et al. 2015). Symptoms of high blood sugar include polydipsia, polyurea and polyphagia and weight loss. Acute complications of diabetes include ketoacidosis, hyperglycemia, or death (Kitabchi et al. 2009). Long-term life-threatening consequences of diabetes include kidney failure, cardiovascular symptoms, foot ulcers, stroke, and loss of vision (WHO 2013). Various approaches for the detection of glucose sensing have been explored. Among all the approaches, FRET-based detection methods are achieving increased attention because of the benefits of molecular fluorescence for biosensing. The FRET-based technique is non-invasive and is extremely sensitive (Weiss 1999). Fluorescence measurements trigger light or no harm to the cells or tissues and penetrate several centimeters of tissue (Lakowicz 1994; Pickup et al. 2005b). By using the phenomenon of FRET, the distribution and structure of biomolecules can be examined (Selvin 1995; Lakowicz 1999). FRET-based nanosensors have been developed for the detection of glucose. To observe glucose level changes in individual living cells, a glucose nanosensor (FLIPglu) was developed by joining the *E. coli* periplasmic glucose/galactose-binding protein along with two GFP derivatives.

Binding of glucose to the FLIPglu-170n version sensor exhibited a concentration-dependent decline in FRET between the attached fluorophores. An affinity mutant of FLIPglu was developed, which showed more substrate specificity as compared to other monosaccharides and thus allowed particular detection of the reversible change in glucose level in COS-7 cells in the physiological range. The FLIPglu offers a novel way with various scientific, environmental, and medical applications.

8.3.3.2 Concanavalin A (ConA)

ConA is a plant carbohydrate-binding protein, which has four binding sites for glucose (Reeke et al. 1975). ConA binds to the glucose and other derivatives of carbohydrates such as glycated protein and dextran on the basis of competitive binding. High molecular-based dextran, which was labeled with fluorescein isothiocyanate (FITC) was the competitive ligand of ConA, in the lumen, the fluorescence intensity increased when the glucose from the external medium replace dextran from ConA (Meadows and Schultz 1988). In the same laboratory, this concept was later used for FRET-based measurement, in which FITC dextran acts as a donor and rhodamine-ConA acts as an acceptor. When the glucose the added, the FRET intensity decrease, and the fluorescence of fluorescein donor rises (Pickup et al. 2005a). Meadows and Schultz (1988) describe the disadvantages that ConA aggregation to form precipitates affects the stability of this sensor. Later on, McCartney et al. (2001) and others also noted the difficulties with this sensor.

8.3.4 FRET-Based Probes for Detection of Alzheimer

Reactive Oxygen Species (ROS) are chemically reactive molecules that are produced due to partial reduction of molecular oxygen (Del Rio 2015). Free radicals include superoxide radicals, hydroxyl radicals, hydroperoxyl radicals, alkoxyl radicals or non-radicals include singlet oxygen and hydrogen peroxide which are formed from reactive oxygen molecules due to the presence of one or more unpaired electrons. ROS, which are produced endogenously, acts as a signaling molecule in certain biological processes but excessive production of ROS causes cellular damage and oxidative stress which results in peroxidation of lipids, proteins denaturation, degradation of nucleotides, and finally cell death (Ahsan et al. 2003). Superoxide radicals and hydrogen peroxide are the major sources of endogenous ROS, which are produced through cellular respiration (Nohl et al. 2003). Oxidative stress plays a vital role in the formation of a wide variety of diseases which include some neurodegenerative disorders, chronic inflammatory diseases, atherosclerosis, and ischemia (Droge 2002). Globally, among neurodegenerative disorders, Alzheimer's disease (AD) is the major health problem that is distinguished by memory loss, speech, and reasoning. In the population over the age of 85, the rate of Alzheimer's disease reaches to 20–40%. Post-mortem analysis has indicated the presence of amyloid-β comprising plaques and neurofibrillary tangles. During Alzheimer's disease, amyloid-β containing plaques and neurofibrillary tangles can be seen in a large amount as compared to the normal aging of the brain (Milton 2004). While the initial causes for Alzheimer's disease are not known. Autopsy finding of the AD brain has identified the presence of oxidative stress markers like lipid oxidation products, oxidized DNA bases, and protein carbonyls, in addition with amyloid-β containing plaques and neurofibrillary tangle.

Hydrogen peroxide is an uncharged, stable, and freely diffusible ROS which acts as a secondary messenger (Milton 2004). Due to the high oxygen intake in the tissue, the generation of hydrogen peroxide is relatively high in the brain. Traditional approaches for the detection of H_2O_2 in living cells suffer from numerous issues: (1) The probes are not specific to H_2O_2 and reacts with other ROS as well as RNS. (2) H_2O_2 is highly specific in cellular compartments and quickly diffuses across the membranes and often suffer from photostability issues. Boronate-deprotection mechanism-based small-molecule fluorescence probes have been developed for the detection of H_2O_2 (Dickinson et al. 2010, 2011; Miller et al. 2005, 2007; Dickinson and Chang 2008). However, the limitation of small-molecule fluorescent probes is their use in vivo studies. Belousov et al. (2006) developed the Hyper and their mutants which acts as ratiometric fluorescent indicator of H_2O_2 in live cells (Dooley et al. 2004; Hanson et al. 2004; Belousov et al. 2006; Meyer and Dick 2010; Wang et al. 2013). For the construction of this probe, they inserted cpYFP into the regulatory domain of OxyR, a bacterial transcription factor. In vitro, Hyper is able to detect the nanomolar to micromolar concentration of H_2O_2. HyPer-2 and HyPer-3 were also developed to increase the dynamic range of half-oxidation and half-reduction responses which shows a high dynamic range.

8.3.5 FRET-Based Sensor for Nitric Oxide [NO]

Nitric oxide (NO), synthesis occurs from the oxidation of single guanidino nitrogen atom of L-arginine, this phenomenon involves the reduction of molecular oxygen and oxidation of nicotinamide adenosine dinucleotide phosphate. For the production of NO, Flavin adenine dinucleotide and flavin mononucleotide both act as cofactors (Vincent et al. 2000). The formation of nitric oxide occurs by numerous cell types and having both the harmful and positive effects at the cellular and vascular levels. In vitro, NO exerts various cellular effects which include alteration of protein and phospholipid of the membrane, nuclear damage, and inhibits the mitochondrial respiration in numerous cell types. The toxicity effect of NO increases due to the reaction of NO with superoxide radicals (Szabo et al. 1997). Hence excess production of NO or peroxynitrite inside the body causes multiple organ dysfunction that often causes sepsis. In contrast, NO acts as a scavenger for oxygen-free radicals and guard the cells from oxidative damage (Wink et al. 1993) and also inhibits the production of oxygen-free radicals (Clancy et al. 1992). In normal lungs, the maintenance of a low resting pulmonary vascular tone occurs by the basal NO released by cNOS but in sepsis, NO increases pulmonary hypertension and pulmonary vasoconstriction which further increases pulmonary artery pressure due to the imbalance between NO and the strong vasoconstricting endothelins and thromboxane (Petros et al. 1994). Pearce et al. developed the FRET-based sensor for NO detection by using metallothionein (MT), a cysteine-rich metal-binding protein (Palmiter 1998). They choose this protein for the construction of NO sensor because NO induces conformational changes in this protein and in vitro when NO reacts with MT (Kennedy et al. 1993) and the sensitivity of cells to toxic level of NO decreases (Schwarz et al. 1995). This sensor was constructed by sandwiching the cDNA of human MT IIa (hMTIIa) between the cDNA of two mutants of Green fluorescent proteinase CFP and eYFP. The ECFP-hMTIIa-EYFP was digested and then subcloned into the mammalian expression vector pcDNA3.1. It was reported that when NO binds to the MT, changes occur in the intramolecular proximity and relative orientation of the fluorophores (Dos Remedios and Moens 1995) due to the release of metal ions from MT protein and the FRET efficiency F_{535nm}/F_{480nm} decreased (Pearce et al. 2000).

8.3.6 FRET-Based Sensor for Detection of *Mycobacterium tuberculosis*

Tuberculosis (TB) is one of the most severe and widespread infectious disease caused by the bacterium *Mycobacterium tuberculosis* and indicates a global health concern (Yeo et al. 2009). Under tropical field conditions, due to the limited suitability of presently accessible techniques diagnosis of TB even in the developing world is very challenging. Using sensitive culture methods, detection of *Mycobacterium tuberculosis* takes around 6–8 weeks due to the slow growth of this *Mycobacterium*. Over the previous few decades, a lot of work has been done to develop a cost-effective technique (Srivastava et al. 2016). In contrast with the cultivation method, the newly developed FRET-based biosensor showed high sensitivity and specificity. The cost of PCR-based techniques for the detection of *Mycobacterium tuberculosis* is slightly lower as compared to the sandwich-form FRET-based biosensor, but the large scale usage of FRET-based biosensor would be extra cost-effective due to its accuracy and speed (Shojaei et al. 2014).

8.3.7 Probes for the Detection of Superoxide Radicals

In living systems during several physiological and pathophysiological activities, the oxygen-free radicals are known to perform a key role in certain cellular metabolism (Southorn and Powis 1988) but on the other hand, these are highly toxic for the organism because they can alter the biological molecules though oxidation (Saltman 1989). These free radicals are thought to be involved in the tissue injury formed when blood supply returns to tissue after ischemia or lack of oxygen (Southorn and Powis 1988). Due to the oxidative attack, the structural and physical properties of human red blood cells altered which may lead to changes in RBC rheologic behavior. Baskurt et al. were used two different models to detect the effect of superoxide anions on RBCs internally or externally. Their results show that the deformability of RBC decreases but there was no effect seen on cell aggregation through the production of superoxide inside the RBC by phenazine methosulfate. Under normal conditions, in the exceedingly of the RBC, the free radicals are continuously generated but their deleterious impact on cellular components prevented by the antioxidant defense mechanism present in the cell. But due to the imbalance between the antioxidant defense and increased generation of oxygen-free radicals, these oxidatively damaged the RBC by both functional and structural impairments. A novel fluorophore MitoSOX Red was developed for the superoxide detection in human coronary artery endothelial cells and in the mitochondria of rat, cardiac derived H9c2 myocytes (Mukhopadhyay et al. 2007). Ethidium-based probes for fluorescence imaging of superoxide in cultured cells was also developed (Robinson et al. 2006). Schwarzländer et al. have been developed the superoxide probes for the detection of superoxide in mitochondria of plant cells and in cardiomyocytes. This probe was constructed by combining the cpYFP to the mitochondrial-targeting sequence from *Nicotiana plumbaginifolia* *β*-ATPase under the control of CaMV 35S promoter in pH2GW7 vector by using gateway technology (Schwarzländer et al. 2011).

8.4 Conclusions

Due to methodological problems in analyzing the lively distribution of metabolites and/or measurement of metabolite concentrations is limited mainly to the evaluation of average concentrations over entire tissues or organs. Noninvasive monitoring techniques would have major advantages over the other methods in offering an advance understanding of metabolite transport, sensing, and the metabolic reactions compartmentalization. Nanosensors, the noninvasive tools that report processes in living entities or in their environment in real time, to allow us the quantification of the changes of ions, signaling molecules and metabolites are currently on the horizon. Many of the periplasmic binding proteins(PBPs) has been successfully converted into nanosensors for a wide range of metabolites. The designing of these nanosensors is based on the conformational alterations of ligand-binding proteins upon the binding of ligand and the phenomenon of FRET between the fluorescent protein pair. Our understanding of conditions for expressing these nanosensors in pro- and eukaryotes and about the likely caveats is rapidly increasing and would help to decipher the complicated metabolic network in the future. In brief, the sensors were used to determine the cellular and subcellular levels of various analytes, non-invasively. These FRET-based nanosensors may offer a tool that might benefit toward a better knowledge of the mechanisms of noninvasive sensing and metabolic disorders linked with the metabolite homeostasis in all classes of organisms.

Acknowledgment First Author, Amreen is thankful to the Council of Scientific and Industrial Research, Govt. of India (0s9/112(0595)2018EMR-1) for Junior Research Fellowship.

References

Ahsan H, Ali A, Ali R (2003) Oxygen free radicals and systemic autoimmunity. Clin Exp Immunol 131:398–404

Arlt K, Brandt S, Kehr J (2001) Amino acid analysis in five pooled single plant cell samples using capillary electrophoresis coupled to laser-induced fluorescence detection. J Chromatogr 926:319–325

Baird GS, Zacharias DA, Tsien RY (2000) Biochemistry, mutagenesis, and oligomerization of dsRED, a red fluorescent protein from coral. Proc Natl Acad Sci U S A 97:11984–11989

Belousov VV, Fradkov AF, Lukyanov KA, Staroverov DB, Shakhbazov KS, Terskikh AV, Lukyanov S (2006) Genetically encoded fluorescent indicator for intracellular hydrogen peroxide. Nat Methods 3:281

Berger EA (1973) Different mechanisms of energy coupling for the active transport of proline and glutamine in Escherichia coli. Proc Natl Acad Sci U S A 70:1514–1518

Berger EA, Heppel LA (1974) Different mechanisms of energy coupling for the shock-sensitive and shock-resistant amino acid permeases of Escherichia coli. J Biol Chem 249:7747–7755

Bogner M, Ludewig U (2007) Visualization of arginine influx into plant cells using a specific FRET-sensor. J Fluoresc 17:350–360

Bray F, Ferlay J, Soerjomataram I, Siegel RL, Torre LA, Jemal A (2018) Global cancer statistics 2018: GLOBOCAN estimates of incidence and mortality worldwide for 36 cancers in 185 countries. CA Cancer J Clin 68:394–424

Clancy RM, Leszczynska-Piziak J, Abramson SB (1992) Nitric oxide, an endothelial cell relaxation factor, inhibits neutrophil superoxide anion production via a direct action on the NADPH oxidase. Clin Drug Investig 90:1116–1121

Del Rio LA (2015) ROS and RNS in plant physiology: an overview. J Exp Bot 66:2827–2837

Dickinson BC, Chang CJ (2008) A targetable fluorescent probe for imaging hydrogen peroxide in the mitochondria of living cells. J Am Chem Soc 130:9638–9639

Dickinson BC, Huynh C, Chang CJ (2010) A palette of fluorescent probes with varying emission colors for imaging hydrogen peroxide signaling in living cells. J Am Chem Soc 132:5906–5915

Dickinson BC, Peltier J, Stone D, Schaffer DV, Chang CJ (2011) Nox2 redox signaling maintains essential cell populations in the brain. Nat Chem Biol 7:106

Dooley CT, Dore TM, Hanson GT, Jackson WC, Remington SJ, Tsien RY (2004) Imaging dynamic redox changes in mammalian cells with green fluorescent protein indicators. J Biol Chem 279:22284–22293

Dos Remedios CG, Moens PD (1995) Fluorescence resonance energy transfer spectroscopy is a reliable "ruler" for measuring structural changes in proteins: dispelling the problem of the unknown orientation factor. J Struct Biol 115:175–185

Dos Remedios CG, Miki M, Barden JA (1987) Fluorescence resonance energy transfer measurements of distances in actin and myosin: a critical evaluation. J Muscle Res Cell Motil 8:97–117

Droge W (2002) Free radicals in the physiological control of cell function. Physiol Rev 82:47–95

Ellis JM, Wolfgang MJ (2012) A genetically encoded metabolite sensor for malonyl-CoA. Chem Biol 19:1333–1339

Fehr M, Lalonde S, Lager I, Frommer WB (2003) In vivo imaging of the dynamics of glucose uptake in the cytosol of COS-7 cells by fluorescent nanosensors. J Biol Chem 278:19127–19133

Feller SM (2001) Crk family adaptors–signalling complex formation and biological roles. Oncogene 20:6348

Fitzmaurice C, Akinyemiju TF, Al Lami FH, Alam T, Alizadeh-Navaei R, Allen C, Alsharif U, Alvis-Guzman N, Amini E, Anderson BO, Aremu O (2018) Global, regional, and national cancer incidence, mortality, years of life lost, years lived with disability, and disability-adjusted life-years for 29 cancer groups, 1990 to 2016: a systematic analysis for the global burden of disease study. JAMA Oncol 4:1553–1568

Forster T (1965) Delocalization excitation and excitation transfer. In: Modern quantum chemistry. Academic, New York, pp 93–137

Funfschilling U, Supplie LM, Mahad D, Boretius S, Saab AS et al (2012) Glycolytic oligodendrocytes maintain myelin and long-term axonal integrity. Nature 485:517–521

Ganapathy V, Thangaraju M, Prasad PD (2009) Nutrient transporters in cancer: relevance to Warburg hypothesis and beyond. Pharmacol Ther 121:29–40

Groffen J, Stephenson JR, Heisterkamp N, de Klein A, Bartram CR, Grosveld G (1984) Philadelphia chromosomal breakpoints are clustered within a limited region, bcr, on chromosome 22. Cell 36:93–99

Hanson GT, Aggeler R, Oglesbee D, Cannon M, Capaldi RA, Tsien RY, Remington SJ (2004) Investigating mitochondrial redox potential with redox-sensitive green fluorescent protein indicators. J Biol Chem 279:13044–13053

Heisterkamp N, Groffen J, Stephenson JR et al (1982) Chromosomal localization of human cellular homologues of two viral oncogenes. Nature 299:747–749

Herman JG, Baylin SB (2003) Gene silencing in cancer in association with promoter hypermethylation. N Engl J Med 349:2042–2054

Imamura H, Nhat KPH, Togawa H, Saito K, Iino R, Kato-Yamada Y, Nagai T, Noji H (2009) Visualization of ATP levels inside single living cells with fluorescence resonance energy transfer-based genetically encoded indicators. Proc Natl Acad Sci U S A 106:15651–15656

Jones AM, Danielson JÅ, ManojKumar SN, Lanquar V, Grossmann G, Frommer WB (2014) Abscisic acid dynamics in roots detected with genetically encoded FRET sensors. elife 3:e01741

Kaper T, Looger LL, Takanaga H, Platten M, Steinman L, Frommer WB (2007) Nanosensor detection of an immunoregulatory tryptophan influx/kynurenine efflux cycle. PLoS Biol 5:2201–2210

Kennedy MC, Gan T, Antholine WE, Petering DH (1993) Metallothionein reacts with Fe^{2+} and NO to form products with ag = 2.039 ESR signal. Biochem Biophys Res Commun 196:632–635

Kitabchi AE, Umpierrez GE, Miles JM, Fisher JN (2009) Hyperglycemic crises in adult patients with diabetes. Diabetes Care 3:1335–1343

Kurzrock R, Gutterman JU, Talpaz M (1988) The molecular genetics of Philadelphia chromosome–positive leukaemia's. New Engl J Med 319:990–998

Lager I, Looger LL, Hilpert M, Lalonde S, Frommer WB (2006) Conversion of a putative Agrobacterium sugar-binding protein into a FRET-sensor with high selectivity for sucrose. J Biol Chem 281:30875–30883

Lakowicz JR (1994) Emerging biomedical applications of time-resolved fluorescence spectroscopy. In: Topics in fluorescence spectroscopy. Springer, Boston, pp 1–19

Lakowicz JR (1999) Fluorescence sensing. In: Principles of fluorescence spectroscopy. Springer, Boston, pp 531–572

Lee Y, Morrison BM, Li Y, Lengacher S, Farah MH, Hoffman PN, Liu Y, Tsingalia A, Jin L, Zhang PW, Pellerin L (2012) Oligodendroglia metabolically support axons and contribute to neurodegeneration. Nature 487:443

Lee T, Cai LX, Lelyveld VS, Hai A, Jasanoff A (2014) Molecular level functional magnetic resonance imaging of dopaminergic signaling. Science 344:533–535

Lesmana J, Friedl P (2001) Destabilization of green fluorescent protein by substitution of its amino-terminal residue. In: Animal cell technology: from target to market. Springer, Dordrecht, pp 6–9

Mao B, Pear MR, McCammon JA, Quiocho FA (1982) Hinge-bending in L-arabinose-binding protein. The "Venus's-flytrap" model. J Biol Chem 257:1131–1133

Masharina A, Reymond L, Maurel D, Umezawa K, Johnsson K (2012) A fluorescent sensor for GABA and synthetic GABAB receptor ligands. J Am Chem Soc 134:19026–19034

McCartney LJ, Pickup JC, Rolinski OJ, Birch DJ (2001) Near-infrared fluorescence lifetime assay for serum glucose based on allophycocyanin-labeled concanavalin A. Anal Biochem 292:216–221

Meadows D, Schultz JS (1988) Fiber-optic biosensors based on fluorescence energy transfer. Talanta 35:145–150

Meyer AJ, Dick TP (2010) Fluorescent protein-based redox probes. Antioxid Redox Signal 13:621–650
Meyer-Lindenberg A (2010) From maps to mechanisms through neuroimaging of schizophrenia. Nature 468:194–202
Miller EW, Albers AE, Pralle A, Isacoff EY, Chang CJ (2005) Boronate-based fluorescent probes for imaging cellular hydrogen peroxide. J Am Chem Soc 127:16652–16659
Miller EW, Tulyathan O, Isacoff EY, Chang CJ (2007) Molecular imaging of hydrogen peroxide produced for cell signaling. Nat Chem Biol 3:263
Milton NG (2004) Role of hydrogen peroxide in the aetiology of Alzheimer's disease. Drugs Aging 21:81–100
Miyawaki A, Llopis J, Heim R, McCaffery JM, Adams JA, Ikura M, Tsien RY (1997) Fluorescent indicators for $Ca2^+$ based on green fluorescent proteins and calmodulin. Nature 388:882–887
Mizutani T, Kondo T, Darmanin S, Tsuda M, Tanaka S, Tobiume M, Asaka M, Ohba Y (2010) A novel FRET-based biosensor for the measurement of BCR-ABL activity and its response to drugs in living cells. Clin Cancer Res 16:3964–3975
Mohsin M, Abdin MZ, Nischal L, Kardam H, Ahmad A (2013) Genetically encoded FRET-based nanosensor for in vivo measurement of leucine. Biosens Bioelectron 50:72–77
Mukhopadhyay P, Rajesh M, Yoshihiro K, Haskó G, Pacher P (2007) Simple quantitative detection of mitochondrial superoxide production in live cells. Biochem Biophys Res Commun 358:203–208
Muller A, Joseph V, Slesinger PA, Kleinfeld D (2014) Cell-based reporters reveal in vivo dynamics of dopamine and norepinephrine release in murine cortex. Nat Methods 11:1245
Nagai T, Sawano A, Park ES, Miyawaki A (2001) Circularly permuted green fluorescent proteins engineered to sense Ca^{2+}. Proc Natl Acad Sci U S A 98:3197–3202
Newman LA, Korol DL, Gold PE (2011) Lactate produced by glycogenolysis in astrocytes regulates memory processing. PLoS One 6:e28427
Nohl H, Kozlov AV, Gille L, Staniek K (2003) Cell respiration and formation of reactive oxygen species. Facts and artefacts. Biochem Soc Trans 31:1308–1311
Okumoto S, Looger LL, Micheva KD, Reimer RJ, Smith SJ, Frommer WB (2005) Detection of glutamate release from neurons by genetically encoded surface-displayed FRET nanosensors. Proc Natl Acad Sci U S A 102:8740–8745
Palmiter RD (1998) The elusive function of metallothioneins. Proc Natl Acad Sci U S A 95:8428–8430
Payne SR, Kemp CJ (2005) Tumour suppressor genetics. Carcinogenesis 26:2031–2045
Pearce LL, Gandley RE, Han W, Wasserloos K, Stitt M, Kanai AJ, McLaughlin MK, Pitt BR, Levitan ES (2000) Role of metallothionein in nitric oxide signaling as revealed by a green fluorescent fusion protein. Proc Natl Acad Sci U S A 97:477–482
Petros A, Lamb G, Leone A, Moncada S, Bennett D, Vallance P (1994) Effects of a nitric oxide synthase inhibitor in humans with septic shock. Cardiovasc Res 28:34–39
Pickup JC, Hussain F, Evans ND, Rolinski OJ, Birch DJ (2005a) Fluorescence-based glucose sensors. Biosens Bioelectron 20:2555–2565
Pickup JC, Hussain F, Evans ND, Sachedina N (2005b) In vivo glucose monitoring: the clinical reality and the promise. Biosens Bioelectron 20:1897–1902
Piero MN, Nzaro GM, Njagi JM (2015) Diabetes mellitus-a devastating metabolic disorder. Asian J Pharm 5:1
Polo E, Kruss S (2016) Nanosensors for neurotransmitters. Anal Bioanal 408:2727–2741
Quiocho FA, Ledvina PS (1996) Atomic structure and specificity of bacterial periplasmic receptors for active transport and chemotaxis: variation of common themes. Mol Microbiol 20:17–25
Quiocho FA, Phillips GN Jr, Parsons RG, Hogg RW (1974) Crystallographic data of an L-arabinose-binding protein from Escherichia coli. J Mol Biol 86:491–493
Reeke GN, Becker JW, Edelman GM (1975) The covalent and three-dimensional structure of concanavalin A. IV. Atomic coordinates, hydrogen bonding, and quaternary structure. J Biol Chem 250:1525–1547

Robinson KM, Janes MS, Pehar M, Monette JS, Ross MF, Hagen TM, Murphy MP, Beckman JS (2006) Selective fluorescent imaging of superoxide in vivo using ethidium-based probes. Proc Natl Acad Sci U S A 103:15038–15043

Rusakov DA (2018) An optical sensor to monitor dynamics of extracellular glycine. Nat Chem Biol 14:835

Saltman P (1989) Oxidative stress: a radical view. Semin Hematol 26:249

San Martin A, Ceballo S, Ruminot I, Lerchundi R, Frommer WB, Barros LF (2013) A genetically encoded FRET lactate sensor and its use to detect the Warburg effect in single cancer cells. PLoS One 8:e57712

Schwarz MA, Lazo JS, Yalowich JC, Allen WP, Whitmore M, Begonia HA, Zeng E, Billiard TR, Robbins PD, Lancaster JR (1995) Metallothionein protects against the cytotoxic and DNA-damaging effects of nitric oxide. Proc Natl Acad Sci U S A 92:4452–4456

Schwarzländer M, Logan DC, Fricker MD, Sweetlove LJ (2011) The circularly permuted yellow fluorescent protein cpYFP that has been used as a superoxide probe is highly responsive to pH but not superoxide in mitochondria: implications for the existence of superoxide 'flashes'. Biochem J 437:381–387

Selvin PR (1995) Fluorescence resonance energy transfer. In: Methods in enzymology, vol 246. Academic, New York, pp 300–334

Seong J, Ouyang M, Kim T, Sun J, Wen PC, Lu S, Zhuo Y, Llewellyn NM, Schlaepfer DD, Guan JL, Chien S (2011) Detection of focal adhesion kinase activation at membrane microdomains by fluorescence resonance energy transfer. Nat Commun 2:406

Shimizu H, Watanabe E, Hiyama TY, Nagakura A, Fujikawa A, Okado H, Yanagawa Y, Obata K, Noda M (2007) Glial Nax channels control lactate signaling to neurons for brain [Na+] sensing. Neuron 54:59–72

Shojaei TR, Salleh MA, Tabatabaei M, Ekrami A, Motallebi R, Rahmani-Cherati T, Hajalilou A, Jorfi R (2014) Development of sandwich-form biosensor to detect Mycobacterium tuberculosis complex in clinical sputum specimens. Braz J Infect Dis 18:600–608

Song Y, Zhang D, Polizzi KM, Zhang EG, Li G, Huang WE (2012) Trends of bacterial whole-cell biosensors in environmental health. In: Preedy VR, Patel VB (eds) Biosensors and environmental health. CRC Press, Boca Raton, pp 127–148

Song Y, Yang M, Wegner SV, Zhao J, Zhu R, Wu Y, He C, Chen PR (2015) A genetically encoded FRET sensor for intracellular heme. ACS Chem Biol 10:1610–1615

Southorn PA, Powis G (1988) Free radicals in medicine. I. Chemical nature and biologic reactions. Mayo Clin Proc 63:381–389

Srivastava SK, Van Rijn CJ, Jongsma MA (2016) Biosensor-based detection of tuberculosis. RSC Adv 6:17759–17771

Stawarski M, Rutkowska-Wlodarczyk I, Zeug A, Bijata M, Madej H, Kaczmarek L, Wlodarczyk J (2014) Genetically encoded FRET-based biosensor for imaging MMP-9 activity. Biomaterials 35:1402–1410

Suzuki A, Stern SA, Bozdagi O, Huntley GW, Walker RH, Magistretti PJ, Alberini CM (2011) Astrocyte-neuron lactate transport is required for long-term memory formation. Cell 144:810–823

Szabo C, Cuzzocrea S, Zingarelli B, O'Connor M, Salzman AL (1997) Endothelial dysfunction in a rat model of endotoxic shock: importance of the activation of poly (ADP-ribose) synthetase by peroxynitrite. J Clin Invest 100:723–735

Tam RO, Saier MH (1993) Structural, functional, and evolutionary relationships among extracellular solute-binding receptors of bacteria. Microbiol Mol Biol Rev 57:320–346

Tennant DA, Durán RV, Gottlieb E (2010) Targeting metabolic transformation for cancer therapy. Nat Rev Cancer 10:267

Tsien RY (1998) The green fluorescent protein. Annu Rev Biochem 67:509–544

Vander Heiden MG, Cantley LC, Thompson CB (2009) Understanding the Warburg effect: the metabolic requirements of cell proliferation. Science 324:1029–1033

Vincent JL, Zhang H, Szabo C, Preiser JC (2000) Effects of nitric oxide in septic shock. Am J Respir Crit Care Med 161:1781–1785

Wang XH, Fang HQ, Huang ZL, Shang W, Hou TT, Cheng AW, Cheng HP (2013) Imaging ROS signaling in cells and animals. J Mol Med 91:917–927

Weiss S (1999) Fluorescence spectroscopy of single biomolecules. Science 283:1676–1683

WHO (2013) Diabetes fact sheet No. 312. Archived from the original on 26 August 2013. Retrieved 25 March 2014

Wink DA, Hanbauer I, Krishna MC, DeGraff W, Gamson J, Mitchell JB (1993) Nitric oxide protects against cellular damage and cytotoxicity from reactive oxygen species. Proc Natl Acad Sci U S A 90:9813–9817

Yeo WH, Liu S, Chung JH, Liu Y, Lee KH (2009) Rapid detection of Mycobacterium tuberculosis cells by using microtip-based immunoassay. Anal Bioanal Chem 393:1593–1600

Zhang J, Xing B, Song J, Zhang F, Nie C, Jiao L, Liu L, Lv F, Wang S (2013) Associated analysis of DNA methylation for cancer detection using CCP-based FRET technique. Anal Chem 86:346–350

Application of Nanosensors in Agriculture and Food Processing

9

Zeenat Mushtaq and Shahla Faizan

Abstract

Nanotechnology has arisen as a blessing to the world with enormous prospective in diverse range of study and daily life. The presentation of nanotechnology for the progression of biosensor leads to a capable nanobiosensor with minute structure in comparison to conventional biosensors. Nanobiosensors can be excellently used for recognizing an extensive variety of fertilizers, herbicide, pesticide, insecticide, pathogens, moisture, and "soil pH." The appropriate and exact use of nanobiosensors can support sustainable agronomy for improving yield production. Even though the scientific studies on the uses of nanotechnology in agronomy are not as much of as years old however the scenarios of nanotechnology in this arena are considerable. The fast improvements in the Nanosciences have a countless impact on agronomic activities and food manufacturing industries. Nanotechnology has a huge ability to suggest keener, robust, worthwhile wrapping constituents, biosensors for the quick recognition of the food pathogens, contaminants as well as further toxins or food additives. It too plays a significant part in developing new group of pesticides with the innocuous carriers, preservation, packing of foodstuffs and food condiments, establishment of natural fiber, exclusion of different toxins from the soil as well as water bodies by means of functionalized nanoparticles and enhancing the shelf life of the vegetables, flowers, and fruits.

Keywords

Nanotechnology · Agriculture · Fertilizers

Z. Mushtaq (✉) · S. Faizan
Environmental Physiology Laboratory, Department of Botany, Aligarh Muslim University, Aligarh, Uttar Pradesh, India

M. Mohsin et al. (eds.), *Nanobiosensors for Agricultural, Medical and Environmental Applications*, https://doi.org/10.1007/978-981-15-8346-9_9

Abbreviations

DEF	Dendrimer Enhanced Filtration
EDCs	Endocrine Disrupting Contaminants
ICT	Information and communications technology
LOFAR	Low-frequency Array
MEMS	Micro Electro Mechanical Systems
PPCPs	Pharmaceutical and personal care products
SSCM	Site-Specific Crop Management
ZVI	Zero Valent Iron

9.1 Introduction

Nanotechnology is also recognized as molecular manufacturing. The dictionary definition of "nano" means small. Nanotechnology or molecular manufacturing is a branch of engineering that is geared towards the design and manufacture of extremely small (molecular dimensions) electronic devices and circuits within 0.1–100 nm (nanometer) dimensions. Nanotechnology manipulates the substance on an atomic, molecular, and supramolecular scale. Nanotechnology is able to construct numerous new things and expedients with enormous kinds of advantages, for example, in nanomedicine, nanoelectronics, biomaterials energy production, and consumer products. The uses of nanomaterials especially for agrarian purposes are used for improving the fertilization process, escalation in harvest via nutrient optimization. The agriculture being the backbone of various emerging nations in which the main portion of their earnings arises from it as well as more than half of people depends on it for their livelihood, nano-materials are used for improvement in the fertilization process, production upsurge via nutrient optimization and lessened the necessities of products used for plant defense that have both beneficial as well as harmful influence on the plant (Huang et al. 2015), use of nanoparticles like nano-silicon particles that increases the ability of water up taking in plants. It revolutionizes the agricultural field by the innovation of new systems such as perfection of agricultural methods, increasing the capability of plants to engross nutrients, competent and exact consumption of active ingredients, recognition as well as control of infections, storing and packing and minimize the nutrient losses. Nanotechnology-derived strategies also play an immense part in the areas of genetics, the formation of drought and pest-resistant crops as well as in plant breeding. Lately, the technique of precision farming centered on minute microelectromechanical systems (MEMS) named "smart dust" is considered as upcoming nanotechnology for agrarian practices. Smart dust consists of sensors, robots, and transponders that work on a wireless computer network and sense light, temperature, vibration, magnetism, or chemicals through radio-frequency identification. Smart dust can be speckled through crop field and connected to prevailing agricultural apparatus used in precision agriculture and to a personal computer. For example, ASTRON, the Netherlands Institute for Radio Astronomy, has established a novel radio telescope of the LOFAR (Low-Frequency Array), which has thousands of antennas that are linked to each other with a huge ICT infrastructure. LOFAR Agro has applied it for calculating the microclimate in potato crops and to fight against phytophthora infection in the potato crop. Nanotechnology helps in the reduction of environmental pollution by forming the pesticides, weedicides, and other chemical fertilizers with the help of nanoparticles and nanocapsules that have the ability to overcome or control the pests, weeds, etc., with very lower doses.

Nanobiosensors are the modified version of biosensor that is a compressed analytical tool or device integrating a biologically derived sensitized element that is interconnected to a physiochemical transducer (Turner 2000). A typical nano-biosensor consist of three components; (1) biologically sensitized elements (probe) which comprises biologically formed substances such as receptors, enzymes, antibodies, nucleic acids, molecular imprints, lectins, tissue, microorganisms, organelles, etc. or are bio-mimic constituent that obtains signals from the interested sample and then sends it to transducer. (2) Transducer, which is another component of nano-biosensor acts as an interface and calculates the physical changes that occur with the reaction at the bio-receptor/sensitive biological element then transforming that energy into measurable electrical output and (3) Detector, it traps the signals from the transducer, which are then passed to a microprocessor where they are amplified, analyzed and the data is then transferred to user friendly output and displayed/stored (Shana and Rogers 1994; Cavalcanti et al. 2008). The first biosensor was discovered in the year 1976, which headed to the development of the number of altered biosensors (Updike and Hicks 1967).

9.1.1 Nano- Biosensors and Agriculture

Nowadays, nanobiosensors such as nanodevices, nanocapsules, etc. reveal mesmerizing projections over traditional biosensors. Nanobiosensors have striking benefits, for example, improved detection sensitivity, specificity and have pronounced benefits in diverse arenas including conservational and bioprocess control, detection, and treatment of plant infections, transport of dynamic constituents to the preferred spots (Srilatha 2011), management of wastewater, enrich the nutrient absorption in plants, quality control of food and play a dynamic role in plants and animal breeding's (Prasanna 2007), precision farming techniques, diagnosis of soil diseases which are caused by microbes like (viruses, bacteria, and fungi) with the help of the quantitative measurement of differential oxygen consumption in the respiration of important microbes and harmful ones in the soil. Some of the potential applications of nanobiosensors are listed below (Fig. 9.1):

9.1.2 Control of Pests and Disease Occurrences in Plants

Insects fed on entire types of plants comprising crop plants, grasses, shrubs, woody plants trees, medicinal plants as well as unwanted weeds. Insect pests and diseases

Fig. 9.1 Shows various applications of nanotechnology in the agronomic field

cause huge loss to credible agronomic yield. It was observed that insects and pests cause about 25% loss in rice, 5–10% in wheat, 30% in pulses, 35% in oilseeds, 20% in sugar cane, and 50% in cotton (Dhaliwal et al. 2010). There are numerous nanoparticles used for the formation of pesticides, insecticides, and insect repellants that are very efficaciously used for the regulation of numerous plant diseases initiated by different pathogens (Barik et al. 2008; Owolade et al. 2008; Gajbhiye et al. 2009; Goswami et al. 2010). Silver nanoparticles have robust impeding bactericidal properties and high antimicrobial effect due to its high surface area as well as a great portion of surface atoms. Double capsuled nanosilver particles overcome the fungal disease caused by *Sphaerotheca pannosa var rosae*. This is the very common disease of rose's that is responsible for leaf distortion, crimping of leaves, early defoliation, and also reduces flowering. They are also sprinkled to foliage in order to stop the disease caused by fungi, molds, and rot. Nano alumino-silicate tubes with vigorous constituents are another important nanoparticle. Barik et al. (2008) revised the use of nanosilica as nano-pesticide. The way of control of pests using nanosilica is to depend on the fact that insect pests used a diversity of cuticular lipids for shielding their water barrier and thereby prevent death from dehydration. However, nanosilica has the ability to get engrossed into the cuticular lipids by physisorption and is easily picked up in insect hairs when they are sprayed on plant surfaces and cause the death of insects purely by physical means. Besides it, there are various other nanoparticles like Carbon nanotubes, Titanium dioxide nanoparticles, and Magnetic nanoparticles that are used for various agricultural practices. In the forthcoming period, nanodesigned catalysts will become accessible that can escalate the efficiency of commercially available pesticides as well as insecticides and also decline the dosage level vital for crop plants (Joseph and Morrison 2006). Nanotechnology has tremendous advantages in nanoparticle-mediated gene (DNA) transmission. It can be used to transport genetic material like DNA and other preferred chemical substances into plant tissues for the defense of host plants against pathogen attack (Torney 2009).

9.1.3 Wastewater Treatment and Disinfection by Nanoparticles

At present, the facility of unpolluted and ample freshwater is an utmost challenge encountered by the world for anthropological use as well as agriculture applications (Vorosmarty et al. 2010). Agricultural practices need a large amount of freshwater, but due to overuse of chemical fertilizers, several salts, metals like Cu, Pb, As, etc., water-borne pathogenic microorganisms like cryptosporidium, coliform bacteria, virus, etc., radioactive elements and other agricultural chemicals, groundwater pollution is going to increase with very high pace, which adversely affects the agricultural productivity. To fight against this issue, innovative, sustainable, and low-cost technologies will be required for the management of a large quantity of wastewater produced. Research and development in nanotechnology have facilitated us to find innovative and economically viable resolutions for remediation and purification of this wastewater. Nanoparticle bearing photocatalytic decomposition properties have received approval in cleaning and decontamination of contaminated water (Melemeni et al. 2009). Photocatalysis involves the use of nanoparticles, nanoporous fibers, and nanoporous foams and in this process, the reaction of the nanoparticles with the particular chemical compounds arises in the presence of light. This method can be used for the decomposition of noxious chemical compounds including pesticides that are commonly not tarnished under normal conditions. During the decontamination of water, the nanoparticles, when excited with the light source, negative electrons are released (Li et al. 2009). These electrons can be used for eliminating the microbes and organic contaminants like PPCPs and EDCs from the wastewater.

9.1.4 Precision Farming

Precision agriculture (PA), satellite farming or site specific crop management (SSCM) is agricultural managing perception centered on detecting, calculating, and responding to inter and intra-field variability of crops. The main aim of precision farming is to maximize yield while diminishing the use of fertilizers, pesticides, herbicides through watching environmental variables, and applying targeted action. It involves the use of computers, satellite systems, and remote sensing devices to measure the environmental conditions, provides knowledge about the efficacy of the crop growing, the exact identification of problems, and their nature and location. It also provides proper information to farmers that help them to take better decisions and plans regarding their agricultural practices (Cioffi et al. 2004). Nanoparticles release the fertilizers in very slow speed because of having higher surface tension and thus hold the substantial more powerfully from the plants. They also have the nano coating that diminishes the rate of dissolution of the fertilizers and permits the efficient absorption of coated fertilizers by plants. Nanocoating also provides surface protection to larger particles (Brady et al. 1999; Santoso et al. 1995). This technique is very ecofriendly due to the minimum use of chemical fertilizers, slow-release process, reduction of agricultural wastes, and thus saves the environment from pollution (Wu and Liu 2008).

9.1.5 Diagnostic Tool for Soil Quality, Presence of Contaminants, and Other Molecules in Soil

Microbes in soil cause various types of infection and are responsible for numerous diseases. They disturb the property and quality of the soil. Nanosensors showed the marked advantage in the detection of microorganisms led soil diseases through the help of quantifiable calculation of differential oxygen consumption in the respiration of useful microbes and harmful microbes present in the soil. This is the two-step process, two sensors filled with useful and harmful microorganisms respectively, are dipped in a suspension of the soil sample in buffer solution and the oxygen consumption data by two microbes was detected. After comparing the two data, it is easy to judge which microbe favors the soil and finally the quality of the soil.

A number of nanobiosensors are formed to identify the presence of impurities, pests, nutrient content, and plant abiotic stress lead due to drought, temperature, or pressure. Liposome-based biosensors are used to monitor the organophosphorus pesticides such as dichlorvos and paraoxon when present at very low levels. (Vamvakaki and Chaniotakis 2006). Zhang et al. (2007) established a way for the recognition of *Escherichia coli* (*E. coli*) by means of bismuth Nanofilm modified GCE based on flow injection analysis (FIA) principle. Seo et al. (2008) made a biochip sensor system, involving of two Ti contact pads and a 150 nm wide Ti nanowell device on $LiNbO_3$ substrate. When the bacteria were resistant to the phages (uninfected bacteria), slight power variations were detected in the nanowell showing a power spectral density (PSD). The biosensors established using PSII (photosystem

II), recognized to bind some groups of herbicides, isolated from photosynthetic organisms could have the capability to display contaminating chemicals, prominent to the system of a cost-effective, easy-to-use device, competent to expose particular herbicides, and ultimately, a wide range of organic compounds present in industrial and urban effluents, sewage sludge, landfill leak-water, groundwater, and irrigation water.

9.1.6 Role of Nanosensors in Bioremediation and Recycling of Agricultural Wastes

Nanotechnology has played a prominent character in microbial remediation. Ligand-based nanocoated biosensors having high absorption tendency are used for the active exclusion of heavy metals. It befits inexpensive as it can be rejuvenated in situ by treating by means of bifunctional self-assembling ligand of the earlier used nanocoating media. Several coatings of metal can be attached to the same substrate using crystal clear technologies and this technology is expected to be available in the upcoming future (Farmen 2009). According to (Diallo 2009) one more strategy for the exclusion of heavy metals is the use of dendrimer-enhanced filtration (DEF) and can bind with cations and anions according to acidity. In agricultural practices, the use of some chemicals such as pesticides, weedicides, etc., take a lot of time to degrade or are resilient to degradation in nature therefore, persist in the environment for a extensive period and become a source of dangerous problems, but due to the blessings of nanotechnology, these lethal or dangerous compounds can be tarnished under certain situations. If these substances are not destroyed, they may move in the food chain and can cause severe health issues. Current advancements in agronomic nanotechnology have revealed auspicious step in this direction. For example, nanoparticle-water slurry can be mixed in polluted soil, and in due course of time, these practices will decrease the noxiousness of sluggishly degradable or resilient pesticides.

9.1.7 Nano Based Products

9.1.7.1 Nano Herbicides

Unwanted plants called weeds are the major menace in agriculture. They use the essential available nutrients of crop plants to decline the crop yield to a large extent. Elimination of weeds by conventional ways is time consuming. Numbers of herbicides are available commercially. Their application in field kills the weeds, but also damages crop plants as well as decreases the soil fertility and leads to soil pollution. The use of herbicides for an extended duration of time inconsistently leaves remains in the soil, which causes destruction to subsequent crops and weeds resistance in contrary to the same herbicide (Chinnamuthu and Boopathi 2009). Nanoherbicides shows great advantage by eradicating weeds from crops in an environmentally friendly manner, without leaving any harmful remains in the soil and environment (Pérez-de-Luque and Rubiales 2009). Polymeric nanoparticles used for encapsulation of herbicide are very safe for the environment (Kumar et al. 2015). Target specific nanoparticles encumbered with herbicide has been developed for delivery in roots of weeds. These particles go into the root system of the weeds, translocate to cells, and obstruct metabolic pathways such as glycolysis and eventually lead to the death of plants (Ali et al. 2014; Clemente et al. 2014). Noxiousness of poly (ε-caprolactone) nanocapsules having ametryn and atrazine against alga *Pseudokirchneriellasubcapitata* and the micro crustacean *Daphnia similis* has been tested. Herbicides encapsulated in the poly(ε-caprolactone) nanocapsules resulted in lower toxicity to the alga (*Pseudokirchneriellasubcapitata*) and higher toxicity to the micro crustacean *Daphnia similis* in comparison to the herbicides alone (Shyla et al. 2014).

9.1.7.2 Nano Fungicides

Nanotechnology can play a dynamic role in controlling the fungal diseases of plants. Nanoparticles, for example, zinc oxide (35–45 nm), silver (20–80 nm), and titanium dioxide (85–100 nm) has been tested against *Macrophomina phaseolina*, a major soil-borne pathogen of pulse and oilseed crops due to their antifungal property. The greater antifungal activity was detected in silver nanoparticles at minute quantity than zinc oxide and titanium dioxide nanoparticles (Suriyaprabha et al. 2014). Application of nanosilica (20–40 nm) to maize has been observed to provide resistance against phytopathogen, *Fusarium oxysporum* and *Aspergillus niger* in comparison to bulk of silica. The nanosilica-treated plant showed a higher expression of phenolic compounds (2056 and 743 mg/mL respectively) in collected leaf extracts and low expression of stress-responsive enzymes against these fungi. These results showed significantly higher resistance in maize treated with nanosilica than with bulk in terms of disease index and expression of total phenols, phenylalanine ammonia lyase, peroxidase, and polyphenol oxidase, at 10 and 15 kg/ha. Thus silica nanoparticles can be used as a substitute and effective antifungal agent against phytopathogens (Velmurugan et al. 2009). The silver has considerably greater antifungal property than other metals. The reason is that silver ions create the inactivation of thiol groups of fungal cell wall resulting in the distraction of transmembrane, metabolism of energy, and electron transport chain. Changes in fungal genetic material, disruption of the enzyme complexes that are essential for the respiratory chain, condensed membrane permeability, and cell lysis are also other mechanisms (Pal et al. 2007). The efficacy of silver nanoparticles is reliant on particle size and shape and declines with increasing particle size. It has been found that the truncated triangular particle shape showed a more "cidal" effect than circular and rod-shaped particles (Panáček et al. 2006; Karimi et al. 2012). Well dispersed and sustained silver nanoparticle solution can act as an outstanding fungicide due to good adhesion on the bacterial and fungal cell surface (Kim et al. 2009).

9.1.8 Enhancement of Shelf Life of Agroproducts

The innovation in nanotechnology enables us to develop techniques that help us to preserve the vegetables and fruits for a long period. These preserved food products are safe and do not have any side effects. In nano-food system, packing plays an important part in averting post-production or post-harvest damages, besides extending shelf life of fresh and stored agronomic products. It has been estimated that more than 45% of the food packaging industries globally will apply nanotechnology conceptions to advance the preserving feature of agronomic products by 2016. Till today, about 300 nanotechnologies created packing constituents are being sold globally. Presently, transparent plastic nanofilms produced by Chemical giant Bayer (Leverkusen, Germany) add nanoclays or silver nanoparticles into conventional packaging material to advance the ductile properties, stiffness, dimensional stability, and thermal resistance. Such methods facilitate the expansion of the shelf life of foodstuffs by eluding microbial impurity or shelving the growth of microbes by blocking oxygen, carbon dioxide, and moistness from reaching fresh meats or other food products. Furthermore, the integration of silver nanoparticles in packing constituents will help as an antimicrobial (inhibitory) agent and safeguard the farming products from decomposition instigating microbes. Nowadays, Nanocor, which is a subordinate of Amcol International Corporation (Hoffman Estates, USA), is manufacturing nano-composites for usage in beer bottles made of plastic that provide the beverage up to 6-month shelf life. By inserting nano-crystals in plastic, investigators have made a molecular barricade that prevents the escape of oxygen. Nanocor and Southern Clay Products (k2Austin, USA) are currently working on a plastic beer bottle that can enhance their shelf life up to 18 months. Researchers in the Netherlands are also working to develop intelligent packaging that will discharge a preserving ingredient to the food stuffs within packed bags if begins to spoil. This "release on command" preservative packaging functions by expending a bio switch developed through nanotechnology. "Smart" food packaging will warn when oxygen has got inside, or if food is going off. Such packaging is already in use in brewing and dairy production and consists of nano-filters that can filter out microorganisms and even viruses. Nanocapsules delivered chemicals in rapeseed cooking oil, will stop cholesterol entering the bloodstream. Nano scale filters used in the packaging of milk and water have the ability to remove the microbes them without boiling. Advances in processes for producing nanostructured materials coupled with appropriate formulation strategies have enabled the production and stabilization of nanoparticles that have potential applications in the food and related industries (Sanguansri and Augustin 2006; Bouwmeester and Sips 2007).

9.2 Conclusions

Nanotechnology has the ability to modernize the current technologies used in many areas including agriculture. It may have potent resolutions against numerous agronomy associated issues like insect pest management using traditional methods, adverse effects of chemical pesticides, development of improved crop varieties, etc. Nanomaterial in diverse methods can be used for proficient management of insect pests and formulations of potential insecticides and pesticides. Nanoparticle-mediated gene transfer would be useful for the development of new insect-resistant varieties. Therefore, it can also be concluded that nanotechnology can provide green and ecofriendly alternatives for insect pest management without harming nature.

Acknowledgments The first author is thankful for managing editor for kindly inviting this book chapter to be submitted in their book. The author is also thankful to the University Grants Commission for their financial support in terms of Non-NET Fellowship.

References

Ali MA, Rehman I, Iqbal A, Din S, Rao AQ, Latif A, Samiullah TR, Azam S, Husnain T (2014) Nanotechnology: a new frontier in agriculture. Adv Life Sci 1:129–138

Barik TK, Sahu B, Swain V (2008) Nano-silica from medicine to pest control. Parasitol Res 103:253–258

Bouwmeester H, Sips A (2007) Safety for consumers of application of nanotechnologies and nanoparticles in food production. Toxicol Lett 172:S191

Brady NR, Weil RR, Prentice H (eds) (1999) The nature and properties of soils. Pearson, Columbus, pp 415–473

Cavalcanti A, Shirinzadeh B, Zhang M, Kretly LC (2008) Nanorobot hardware architecture for medical defense. Sensors 8(5):2932–2958

Chinnamuthu C, Boopathi PM (2009) Nanotechnology and agroecosystem. Madras Agric J 96:17–31

Cioffi N, Torsi L, Ditaranto N, Sabbatini L, Zambonin PG, Tantillo G, Ghibelli L, D'Alessio M, Bleve-Zacheo T, Traversa E (2004) Antifungal activity of polymer-based copper nano-composite coatings. Appl Phys Lett 85:2417–2419

Clemente Z, Grillo R, Jonsson M, Santos NZ, Feitosa LO, Lima R (2014) Ecotoxicological evaluation of poly(ε-caprolactone) nanocapsules containing triazine herbicides. J Nanosci Nanotechnol 14:4911–4917

Dhaliwal GS, Jindal V, Dhawan AK (2010) Insect pest problems and crop losses: changing trends. Indian J Ecol 37:1–7

Diallo M (2009) Water treatment by dendrimer-enhanced filtration: principles and applications in nanotechnology applications for clean water. In: Savage N, Diallo M, Duncan J, Street A, Sustich R (eds) . William Andrew Inc., Norwich, p 143. Chapter 11

Farmen L (2009) Commercialization of nanotechnology for removal of heavy metals in drinking. In: Savage N, Diallo M, Duncan J, Street A, Sustich R (eds) water Nanotechnology applications for clean water. William Andrew Inc.). Chapter 9, Norwich, p 115

Gajbhiye M, Kesharwani J, Ingle A, Gade A, Rai M (2009) Fungus mediated synthesis of silver nanoparticles and its activity against pathogenic fungi in combination of fluconazole. Nanomedicine 5:282–286

Goswami A, Roy I, Sengupta S, Debnath N (2010) Novel applications of solid and liquid formulations of nanoparticles against insect pests and pathogens. Thin Solid Films 519:1252–1257. http://www.aist.go.jp/aiste/latestresearch/2004/20040402-1/20040402-1.html

Huang S, Wang L, Liu L, Hou Y, Li L (2015) Nanotechnology in agriculture, livestock, and aquaculture in China. A review. Sustain Dev 35:369–400

Joseph T, Morrison M (2006) Nanotechnology in agriculture and food: a nanoforum report. www.nanoforum.org. Accessed 19 Nov 2011

Karimi N, Minaei S, Almassi M, Shahverdi AR (2012) Application of silver nanoparticles for protection of seeds in different soils. Afr J Agric Res 7:863–1869
Kim SW, Lamsal K, Kim YJ, Kim SB, Jung M, Sim SJ, Kim HS, Chang SJ, Kim JK, Lee YS (2009) An in vitro study of the antifungal effect of silver nanoparticles on oak wilt pathogen *Raffaeleasp*. J Microbiol Biotechnol 19:760–764
Kumar S, Bhanjana G, Sharma A, Sidhu MC, Dilbaghi N (2015) Herbicide loaded carboxymethyl cellulose nanocapsules as potential carrier in Agri nanotechnology. Sci Adv Mater 7:1143–1148
Li Q, Wu P, Shang JK (2009) In: Savage N, Diallo M, Duncan J, Street A, Sustich R (eds) Nanostructured visible-light photocatalysts for water purification nanotechnology applications for clean water. William Andrew Inc.) chapter 2, Norwich, p 17
Melemeni M, Stamatakis D, Xekoukoulotakis NP, Mantzavinos D, Kalogerakis N (2009) Disinfection of municipal wastewater by TiO2 phtocatalysis with UV-A, visible and solar irradiation and BDD electrolysis. Global NEST J 11:357
Owolade OF, Ogunleti DO, Adenekan MO (2008) Titanium dioxide affects disease development and yield of edible cowpea. Elect J Environ Agric Food Chem 7:2942–2947
Pal S, Tak YK, Song JM (2007) Does the antibacterial activity of silver nanoparticles depend on the shape of the nanoparticle? A study of the gram-negative bacterium *Escherichia coli*. Appl Environ Microbiol 73:1712–1720
Panáček A, Kvítek L, Prucek R, Kolář M, Večeřová R, Pizúrová N, Sharma VK, Nevecna T, Zboril R (2006) Silver colloid nanoparticles: synthesis, characterization, and their antibacterial activity. J Phys Chem 110:16248–16253
Pérez-de-Luque A, Rubiales D (2009) Nanotechnology for parasitic plant control. Pest Manag Sci 65:540–545
Prasanna BM (2007) Nanotechnology in agriculture. ICAR National Fellow, Division of Genetics, I.A.R.I., New Delhi, India, pp 111–118
Sanguansri P, Augustin MA (2006) Nanoscale materials development- a food industry perspective. Trends Food Sci Technol 17:547–556
Santoso D, Lefroy RD, Band Blair GJ (1995) Sulfur and phosphorus dynamics in an acid soil/crop system. Aust J Soil Res 33:113–124
Seo S, Dobozi-King M, Young RF, Kish LB, Cheng M (2008) Patterning a nanowell sensor biochip for specific and rapid detection of Bacteria. Microelectron Eng 85:1484–1489
Shana A, Rogers KR (1994) Measurement science and technology. Biosensors 5:461–472
Shyla KK, Natarajan N, Nakkeeran S (2014) Antifungal activity of zinc oxide, silver and titanium dioxide nanoparticles against *Macrophomina phaseolina*. J Mycol Plant Pathol 44:268–273
Srilatha B (2011) Nanotechnology in agriculture. J Nanomedic Nanotechnol 2:123
Suriyaprabha R, Karunakaran G, Kavitha K, Yuvakkumar R, Rajendran V, Kannan N (2014) Application of silica nanoparticles in maize to enhance fungal resistance. IET Nanobiotechnol 8:133–137
Torney F (2009) Nanoparticle mediated plant transformation. Emerging technologies in plant science research. In: Interdepartmental Plant Physiology Major Fall Seminar Series, vol 696. Iowa State University, Iowa
Turner AP (2000) Biosensors-sense and sensitivity. Science 290:1315–1317
Updike SJ, Hicks GP (1967) The enzyme electrode. Nature 214:986–988
Vamvakaki V, Chaniotakis NA (2006) Pesticide detection with a liposome-based nano-biosensor. Biosens Bioelectron 22:2848–2853
Velmurugan N, Gnana Kumar G, Sub Han S (2009) Synthesis and characterization of potential fungicidal silver nano-sized particles and chitosan membrane containing silver particles. Iran Polym J 18:383–392
Vorosmarty CJ, McIntyre PB, Gessner MO, Dudgeon D, Prusevich A, Green P, Glidden S, Bunn SE, Sullivan CA, Liemann CR, Davies PM (2010) Global threats to human water security and river biodiversity. Nature 467:555–561

Wu L, Liu M (2008) Preparation and properties of chitosan coated NPK compound fertilizer with controlled release and water-retention. Carbohydr Polym 72:240–247

Zhang W, Tang H, Geng P, Wang Q, Jin L, Wu Z (2007) Amperometric method for rapid detection of *Escherichia coli* by flow injection analysis using a bismuth nano-film modified glassy carbon electrode. Electrochem Commun 9:833–838

Fluorescent Protein Pairs and Their Application in FRET-Based Nanobiosensors

10

Urooj Fatima and Saher Fatima

Abstract

Convergent evolution of FRET imaging and optical probes have revolutionized the development of new technologies of biological visualization. The advent of fluorescent proteins and FRET have been emerged as a powerful tool to monitor Spatiotemporal resolution of dynamic molecular events in live cells. Different fluorescent proteins (FPs) pairs have been utilized to design the FRET-based biosensors. These FPs are commonly used as donors and acceptors that are genetically encoded and compatible with living cells. In this chapter, we are discussing the palette of FP FRET pairs and FRET-based biosensors as a tool for nanoscale information in biosensing and subsequent analytical application. In addition, a deeper insight into the understanding of the dynamics of biological processes in the living cells will be gained.

Keywords

Fluorescent proteins (FPs) · Fluorescence resonance energy transfer (FRET) · Nanobiosensors · Biosensing

10.1 Introduction

The discovery of green fluorescent protein (GFP) from the jellyfish, *Aequorea victoria* (Shimomura et al. 1962) has emerged from obscurity in the course of bioluminescence studies and opened a new avenue for the exploitation of proteins

U. Fatima (✉)
Department of Botany, Aligarh Muslim University, Aligarh, India

S. Fatima
Department of Agricultural Microbiology, Aligarh Muslim University, Aligarh, India

M. Mohsin et al. (eds.), *Nanobiosensors for Agricultural, Medical and Environmental Applications*, https://doi.org/10.1007/978-981-15-8346-9_10

in molecular and cell biology. The GFP protein is a 28-kDa protein that emits bright green light upon irradiation by ultraviolet (UV) light or blue light (Ward et al. 2005). In the past few decades, fluorescent tools have generated intense interest in biological research and are used as a marker for gene expression and localization to understand the cellular processes by monitoring protein dynamics.

Modulation of fluorescent properties of fluorophores empowers the fluorescent techniques in the study of metabolic processes for the measurements of protein dynamics, stability, turnover, and monitoring of drug delivery (Specht et al. 2017). Different variants of fluorescent protein fusions have revolutionized the imaging technology and enabling quantitative analysis of cellular responses in real time.

The diverse combination of fluorescence proteins for Forster or fluorescence resonance energy transfer (FRET)-based biosensing and bioimaging add sensitivity, reliability, and specificity to become an indispensable tool for the spatial investigation of nano-biological systems (Hotzer et al. 2012).

Moreover, the recent development of an impressive library of fluorophores has been used as FRET pairs in FRET biosensors. These FRET pairs fall into three categories, i.e., fluorescent proteins (FPs), small organic dyes, and quantum dots (QDs). In contrast to dyes and QDs, FPs are genetically encoded and can expand its application in live-cell FRET imaging (Miyawaki et al. 1997). Organic dyes and QDs do not have the ability to label sensing domains without the aid of antibodies (Masharina et al. 2012).

FRET is a unique physical phenomenon where non-radiative energy is transferred via long-range dipole-dipole coupling from a donor (D) molecule to an acceptor molecule (A). It was first described by Theodor Förster in 1946. The rate of energy transfer depends on three parameters: (1) The emission spectrum of the donor must show a significant overlap with the excitation spectrum of an acceptor; (2) Close proximity of a donor and an acceptor fluorophore molecules (within the 1–10 nm range); and (3) The emission and absorption dipole moment of donor and acceptor must be in parallel alignment (Clegg 2009).

FRET-based genetically encoded biosensing tools enable the researchers to interrogate the live cells in a non-destructive and minimally invasive manner. This tool builds by sandwiching ligand sensing domains between a pair of fluorophores. The ligand sensing moiety undergoes a conformational change upon binding with a specific ligand, which brings the pair of fluorophore molecules into close proximity and causes the FRET to occur (Miyawaki 2003). FRET-based biosensors have been adopted as versatile spectroscopic rulers to investigate intracellular dynamic analytes concentrations, protein-protein interactions, and enzyme activities to unravel the understanding of spatiotemporal regulation of various dynamic cellular processes (Miyawaki 2011).

Compared to other technologies such as atomic force microscopy (AFM), surface plasmon resonance (SPR), nuclear magnetic resonance (NMR), and mass spectroscopy (MS), fluorescence-based imaging tool has many advantages because it enables highly sensitive, less-invasive, and safe detection of analytes.

In assembling FRET-based biosensors, selecting donor and acceptor fluorophores as a FRET pairs plays a significant role in the high performance of biosensors in live-cell imaging (Piston and Kremers 2007). Here, we will explore FRET measurement, mechanism of fluorescence, FP FRET pairs, FRET probe design, delivery of the FRET probe into the cell, and applications of FRET-based biosensors.

10.2 Measurement of FRET

FRET can be measured by different spectrofluorometer, fluorescence lifetime imaging, flow cytometry, laser scanning, confocal microscopy, and conventional fluorescence microscopy (Jares-Erijman and Jovin 2003; Sekar and Periasamy 2003). FRET can be detected by the (1) changes in the donor fluorescence; (2) changes in the acceptor fluorescence; (3) changes in donor and acceptor fluorescence concurrently by using spectral imaging; and (4) changes in the position of the fluorophores. However, FRET is a commonly used method for the detection of changes in the donor fluorescence in the presence and absence of acceptor. The fluorescence of the donor molecule should be decreased in the presence of acceptor, indicating the occurrence of FRET. If FRET is occurring, the emission spectra of a donor should get shorter. The analyte concentration is another factor affecting FRET. Excitation filters for the donor should selectively excite the donor molecule, with minimal effect on direct excitation of the acceptor molecule. FRET occurs only when two fluorophores interact with one another within the Foster distance. Negative FRET signals may occur due to the non-parallel alignment of their dipole. The relative emission of the acceptor is measured on donor excitation. FRET approach can be used in living cells by utilizing the photoreversible organic dyes and fluorescent protein (Giordano et al. 2002; Subach et al. 2010). The fluorescence signal from a FRET sample is based on (1) the quenched donor emission; (2) the acceptor emission from direct excitation; (3) the acceptor emission sensitized through the FRET.

10.3 Mechanism of Fluorescence

Fluorescent proteins emit light when they are exposed between blue to ultraviolet-visible light. They absorb light and re-emit it at a higher wavelength. By absorbing photons, electrons in their ground state release energy at a higher wavelength, which led the electrons to jump at a higher orbital. The excitation and relaxation state of electrons can be measured in femtoseconds and picoseconds, respectively. The entire fluorescence lifetime, from excitation to relaxation, occurs instantaneously, and fluorescent emission is only observed while the sample is being illuminated (Haddock et al. 2010). The mechanism of chromophore synthesis is almost similar in all fluorophores, regardless of the source (Barondeau et al. 2003). The mechanism is based on chromophore maturation by posttranslational modification of GFP, which consists of 238 amino acids, of which S65, Y66, and G67 are conserved and form chromophore, which undergoes autocyclization. The fluorophore formation occurs in three steps: (1) Nucleophilic attack of the amide nitrogen atom of Gly67 on the carbonyl carbon of Ser65 resulting in the chromophore formation,

4-(phydroxybenzylidene)-5-imidazolinone, (2) Dehydration of the carbonyl oxygen of Ser65 position, (3) Oxidation of the α–β bond of aromatic amino acid (Tyr66). These modifications yielded a chromophore that, after excitation at either 395 nm or 475 nm absorption maximum, emits green fluorescent light at 509 nm (Tsien 1998). The GFP contains an intrinsic chromophore and does not require any prosthetic group or cofactor for their fluorescence to occur. The absorption and emission of visible light by the chromophore depends on the local environment, subsequent mutations in the chromophore changes the spectra of fluorescent protein which causes an alteration in fluorescence (Heim et al. 1994; Miyawaki et al. 2003). Various GFP-like proteins have been isolated from coral reefs, Anthozoan, and crustaceans and are available commercially. Azami Green is a tetrameric fluorescent protein, a homolog of EGFP isolated from stony coral, Galaxeidae. Because of its tetrameric form, this FP is unavailable for its use as a FRET donor or acceptor and resulting in poor fusion tag. Later, by site-directed and random mutagenesis, this tetramer is converted into monomer and utilized in imaging experiments (Karasawa et al. 2004). The DsRed protein gives fluorescence green initially and then red hue, indicating that this chromophore structure is changed during the maturation process. In this FP, amino acid sequence, i.e., Gln66–Tyr67–Gly68, is analogous to the chromophore-forming sequence in GFP, and form a 4-(phydroxybenzylidene)-5-imidazolinone chromophore similar to GFP, and then generates its green intermediate (Wall et al. 2000). The Cα–Nα bond of Gln66 oxidizes in the maturation process, and protein changes to red color (Gross et al. 2000).

10.4 Types of FP FRET Pairs

10.4.1 CFP-YFP FRET Pairs

The first FP FRET pair was developed by Heim and Tsien (1996), which was made up of green FP (GFP) and blue FP (BFP). Later on, cyan and yellow FP pairs (CFP-YFP) was used to increase brightness by many folds and have greater Forster radius (R_0) values (Kremers et al. 2006). CFP-YFP FRET Pairs improved the quantum efficiency of the donor (CFP) and an extinction coefficient of the acceptor (YFP). However, CFP-YFP pairs suffered from the phototoxicity, photoconversion of YFPs into CFP, and fast photobleaching of YFPs (Lam et al. 2012).

10.4.2 GFP-RFP FRET Pairs

GFP-Red FP pairs have a few limitations of CFP-YFPFRET pairs, such as less phototoxicity and greater spectra separation. TagGFP-TagRFP pairs demonstrated the highest dynamic range in FRET/FLIM analyses in contrast with tested TagGFP-mCherry FRET pairs. When TagGFP-TagRFP pair was compared to TagGFP-mCherry pair, then 1.5-fold increased spectral overlap with increased FRET for the TagGFP-TagRFP pair was reported (Shcherbo et al. 2009).

10.4.3 FFP-IFP FRET Pairs

Monitoring of molecular processes in mammalian tissues via FRET using far-red FPs (FFPs) and infrared fluorescent proteins (IFPs) have been particularly useful for non-invasive in vivo imaging of live cells. As far-red FPs penetrate through hemoglobin rich tissue more rapidly than bluer light so, FFPs-IFPs pairs are helpful in deep tissue in vivo imaging. Chu et al. 2014) used mCardinal, a red-shifted variant for visualization of muscle cell regeneration in live cells and organisms with 633- to 635-nm laser lines. Further development of FFPs with high quantum yield is crucial for in vivo imaging studies.

10.4.4 LSS FP-Based FRET Pairs

Fluorescent protein with a large Stokes shift (LSSFP) is another category of FRET pairs. LSS- FP-Based FRET Pairs fill-up the spectral gap between the yellow and red LSSFPs and are capable of FRET donors for far-red FPs, which allow a larger FRET change in the ratiometric FRET imaging by reducing the spectral crosstalk between the donor and acceptor FPs. It is useful for monitoring several events in a single cell in multicolor FRET imaging. Upon excitation with violet wavelength, LSSYFP mAmetrine gives emission of yellow wavelength and delivered a huge Forster radius, R_0 of 6.6 nm when paired with FP, tdTomato, and has been useful for multicolor FRET. But mAmetrine showed limitations of time-lapse imaging due to poor photostability with a half time of 2.8 min. Later, LSSmOrange, an LSS FP, conferred a five-fold brightness improvement and proved to be an effective FRET donor to mKate2, which is a far-red FPs. LSSmOrange has an excitation and emission at wavelength 437 nm and 572 nm, respectively, which fills the spectral gap between yellow and green FPs, and red LSSFPs (Shcherbakova et al. 2012).

10.4.5 Dark FP-Based FRET Pairs

Dark fluorescent proteins serve as a valuable acceptor for FLIM-FRET. It has a <0.1 smaller quantum yield. Dark FPs are non-fluorescent reduces the risk of artifacts due to spectral contamination and retain their absorption properties, which facilitate FRET to occur. Another features of dark FPs include its signal stability, better folding, and maturation kinetics. In 2015, an improved dark GFP named Shadow G was engineered from previously reported dark fluorescent protein, sREACh, with a major 120-fold reduction in quantum yield. In contrast, dark FPs based sensors like mCherry and sREACh, EGFP-Shadow G reduced spectral contamination and exhibited more stability and precise measurements in voltage, calcium, and Ras sensors (Murakoshi et al. 2015).

10.4.6 Optical Highlighter FP-Based FRET Pairs

Optical highlighter FPs are also well-known as photo transformable FPs (ptFPs). It exhibits unique light-induced photoactivation, photoconversion, and photo-switching properties that are the source of exciting research opportunities (Adam et al., 2014). Optical highlighter FP-based FRET pairs have advantages over regular FPs in FRET because they offer spectral change on the identical samples without the requirement for corrections based on reference images of control cells. Moreover, ptFP-based FRET pairs are independent of photo-destructive procedures that are requisite in apFRET and can make inroads for information on the mobility of proteins when used in the study of protein-protein interaction (Demarco et al. 2006).

10.4.7 Multifluorescent FRET Pairs

The spectral derivatives of FPs have generated a broad range of groups for FRET pair suitable for multicolor FRET, enabling the investigators for the near-simultaneous or simultaneous live-cell imaging of cellular processes with more than two FRET pairs. These multicolor FRET are classified into three categories depending on the excitations, i.e., dual FRET pairs with two excitations, dual FRET pairs with a single excitation, and three FRET pairs and on the number of FRET pairs. It was reported that FRET pairs with two excitations perform well in cells but it lacks to monitor two molecular events simultaneously because of lag time for image acquisition; therefore, it is insufficient for following fast signal dynamics or signal changes in highly motile cells. Thus, a single excitation dual FRET pair was established. Tracing two simultaneous molecular events, cAMP and Ca^{2+} in a neuron, and simultaneous imaging of two FRET sensors based on CFP/YFP and Sapphire/RFP was created. Both donors were excited using a violet light at 405 nm with a little excitation of YFP and RFP acceptors, and four-color images were acquired by placing a quad-channel imager (Niino et al. 2009).

10.5 FRET Probe Design

FRET-based probes depend on two main factors, i.e., a distance and the orientation between the donor-acceptor pair that are essential for spectral overlap between donor emission and acceptor absorption. Zhang et al. (2002) have demonstrated live-cell imaging using the FRET-based probe. GFP found in *Aequorea victoria*, was cloned in 1992 (Prasher et al. 1992) and became the protein tracer in live-cell imaging. The discovery of other FPs from marine organisms and many engineered (Shaner et al. 2004) by creating mutation in GFP different emission profiles have open the door to the application of FRET to study live-cell visualization of cellular events and interaction between molecules. These probes are classified into different categories depending on the approach used for the construction of biosensor to investigate different types of biological processes. There are different strategies of probe design depending upon the approach to study types of cellular events.

10.5.1 Cleavage-Based Approach

In a cleavage-based approach, protease cleavage sequence is flanked by donor and acceptor pair so that the FRET can occur under basal conditions to report the presence of protease. Here, cleavage refers to the disruption of covalent bonds in order to differentiate from unbinding. The presence of protease indicates the binding and digestion of the polypeptide chain linking two FPs that would halt energy transfer and ceases FRET. A similar approach was utilized by Onuki et al. (2002) to monitor the cleavage of an apoptotic facilitator, Bid, by caspase 8. Bid is a member of the Bcl-2 family, which induces cell death. This method was used to report the activity and events related to caspase 8within a live single cell and in real time. In this FRET system, the presence of caspase 8, the excitation of CFP at 433 nm should result in the emission of CFP at 473 nm. Due to the cleavage of peptide bond indicating caspase 8 presence no FRET observe. In the absence of caspase 8, energy should be transferred from excited CFP to YFP, with resulting emission at 525 nm (Onuki et al. 2002). When the fusion protein is cleaved by caspase 8, energy can no longer be transferred from excited CFP to YFP, and no FRET occurs. But protease cleavage is irreversible, so it is not suitable for detecting protease with high basal activity (Fig. 10.1).

10.5.2 Intermolecular FRET Approach

The study of intermolecular protein–protein interaction can be monitored by intermolecular FRET, where one of the protein interacts with the donor and the other to the acceptor. This technique is useful for the investigation of cancer biology and drug development. The type of interaction can be a heteromeric or homomer. This approach has been employed for the detection of intermolecular protein–protein interactions which are important for monitoring many physiologically significant

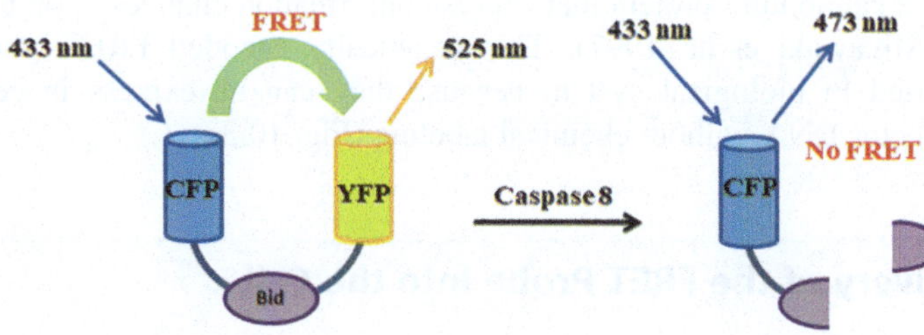

Fig. 10.1 The polypeptide chain remains intact in the absence of protease leading in the FRET to occur while the presence of caspase 8 cleaves the apoptotic facilitator, Bid refers to the disruption of covalent bonds and halt the process of FRET. Adapted from Onuki et al. (2002)

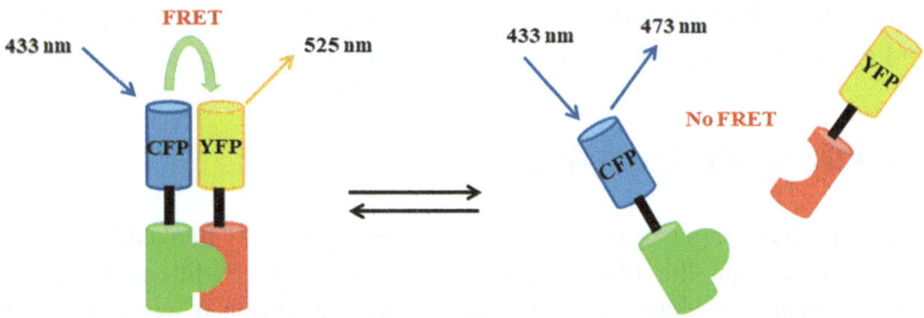

Fig. 10.2 Strategy for detection of Intermolecular heteromer protein interaction. FRET between CFP and YFP represents the interaction between proteins where one party of the protein complex is tagged by a donor and the other by an acceptor. Adapted from Ueda et al. (2013)

cellular processes. Okamoto et al. (2004) studied the dynamic equilibrium between the actin monomer (F-actin/G-actin) to study actin dynamics. In this study, monomer actin was tagged with both donor (CFP) and acceptor (YFP) molecules that would allow the monitoring of dynamic equilibrium between F-actin and G-actin (polymerization/depolymerization) in living cells (Fig. 10.2).

10.5.3 Intramolecular FRET Approach

This approach detects the conformational changes within a given protein when the two probes are localized on a single protein. The conformational changes occur by changes in either the distance or the orientation of the probes that are located on the same molecule. The strategy of two probes attached to a single protein at N- and C-terminus, respectively, outweigh more complex interactions such as the differential redistribution of the donor and acceptor proteins and the expression of two probes at a 1:1 stoichiometry among cells. This approach has been employed for the monitoring of small biological molecules, covalent modifications of proteins, membrane voltage, and signal transduction. This Intramolecular FRET approach was pioneered by Tsien and co-workers for the construction of calcium sensors, cameleons. This cameleon sensors are based on two fluorescent proteins that are attached with a calmodulin protein that causes conformation changes upon binding of calcium (Miyawaki et al. 1997). This genetically encoded FRET-based are extensively used in biological system because they can be express in cells by introducing vector DNA without chemical labeling (Fig. 10.3).

10.6 Delivery of the FRET Probe Into the Cell

Without delivering a FRET probe into a cell, it seems a definite non-starter for an understanding of the detection of intracellular phenomena. Although their introduction across the cell membrane is mechanically introduced by microinjection that

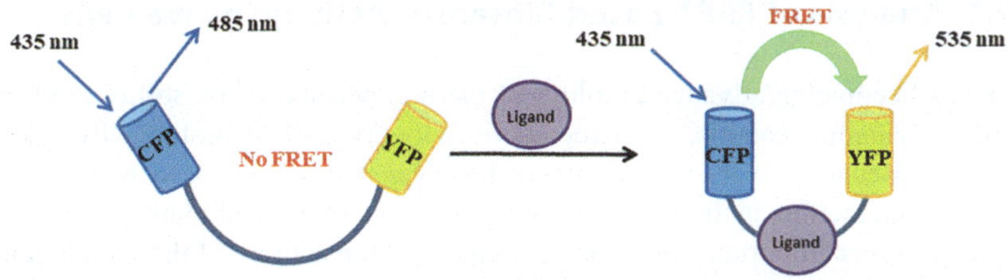

Fig. 10.3 Binding of ligand produces conformational changes in the ligand sensing domain that lead the FRET pairs in close proximity (within a range of 1–10 nm) and spectral overlap, which ultimately causes FRET to occur

creates pore or cell-penetration (Phillip et al. 2012). Transfected cells expressing fluorescent proteins automated about the spatial organization of the target proteins to which they are bound and capable of achieving resolution closer to the molecular scale. In the previous report, a man-made semiconducting nanoscale crystals, quantum dots (QDs) that can transport electrons are synthesized in organic solvents using organometallic precursors and passivated with long, aliphatic trioctylphosphine oxide (TOPO) ligands utilized as donors with FRET for the development of protease sensors for the measurement of extracellular matrix metalloproteinases (MMPs) activity in normal and cancerous breast cells (Shi et al. 2006). In another study, QD-based FRET immunoassays with low picomolar detection limits required the conjugation of antibodies to the ZnS shell of quantum dots that demonstrated the QD-antibody application for nanocrystal-based biosensing (Bhuckory et al. 2016). Toxicity studies become important when a probe is designed for its use in vivo.

In vitro cytotoxicity assay includes the cell line used, incubation time and concentration of analyte, etc. It gives information about whether toxicity presents a significant concern at concentrations comparable to those that might be used in intracellular detection. The introduction of organic dyes results in extremely poor cellular viability (below 10%). FPs utilized for in vivo measurement showed negligible cytotoxicity because of its biological origin (Carpentier et al. 2009). A range of genetically encoded nanosensors have been created by exploiting different FPs pair, introduced non-invasively without the introduction of cofactor or the destruction of the biological sample, and ultimately heralded a new age of cell biology exploration (Frommer et al. 2009; Zhao and Yang 2015; Okumoto and Versaw 2017).

10.7 Applications of FRET-Based Biosensor

FRET-based biosensors are used to analyze different features of protein-protein interactions in the living cells. Moreover, the combination of spectral FRET methods with microscopy offers the unique opportunity to screen macromolecular interactions in living cells and in real time.

10.7.1 Analysis of FRET-Based Biosensor Activity in Live Cells

FRET has become an advanced biological method because of its ability to observe real time dynamic reactions in a metabolic network and in biologically relevant conditions. Because of its sensitivity to distance, it has been used to investigate molecular interaction in living cells. Scrutinization of recombinant proteins in a living cell is very important for the measurement of the activity of these endogenous recombinant proteins. Immunofluorescent staining and green fluorescent protein tagging are suitable for subcellular localization of proteins in-situ within different cell lineages or individual cells (Li-En Jao and Yu Chen 2002). Flow cytometry is another technique for monitoring single cell signaling events (Krutzik and Nolan 2003). To know the conformational changes as well as intracellular species interactions within cells, the resolution limitations of light are not enough but using a highly sensitive and specific assay of molecular proximity indicated by energy transfer is requisite. Detailed information about protein–protein interaction and posttranslational modifications are critical for illustrating the nature and dynamic cellular processes. Thus, FRET probes have been explored for monitoring protein–protein interactions, cellular uptake and unpacking, molecular trafficking and localization processes, and for flux measurement.

10.7.1.1 Protein–Protein Interactions

In the early 1970s, work on FRET for distance measurements on the nanometer scale was established. FRET as a spectroscopic ruler, revealing the spatial proximity relationships of pair of multiple, colored fluorescent proteins labeled in biological macromolecules (Stryer 1978). Increased interest of researchers in FRET techniques has revolutionized the imaging of live cell to examine protein–protein interactions by using Genetic labeling of fluorescent proteins (Frommer et al. 2009; Miyawaki 2011). Live-cell imaging of protein–protein interactions can elucidate individual molecular association, interaction, and binding. The ability to track the activity in living cells increases our understanding of precise regulation. GFP bears excellent properties that enable its use to monitor protein-protein interactions because it does not require cofactor, retains the biochemical functions of the original protein, and can be targeted to specific organelles by adding an appropriate signaling peptide. Most importantly, mutants of GFP created by site-directed mutagenesis show spectral properties and can be used as donors and acceptor fluorophores for FRET to occur (Heim et al. 1994; Delagrave et al. 1995). A notable example emphasizes the variability and proficiencies of the FRET-based technique to monitor protein–protein interactions at subcellular locations. Phillip et al. (2012) monitor the binding of a β-lactamase to β-lactamase inhibitor protein fused with yellow FP in living HeLa cells introduced through microinjection. The addition of an extracellular ligand brought the FPs into proximity and restored FRET between these FPs. In HIV research, Poole et al. (2005) analyzed HIV-1 Gag-RNA Interaction by Confocal Microscopy and FRET that provide new insights into this critical process in HIV replication. Add ref.

10.7.1.2 Intracellular Entry and Unpacking

The delivery of biomolecules into living cells holds a great promise for therapeutic and research applications. FRET has been utilized for dynamic monitoring of the fate of an intracellular targeted molecule as it proceeds through a biochemical pathway and to study the physicochemical properties of nanocomplexes. Cellular uptake and endocytosis are important processes for physiologically and pharmacologically study. A rapid advancement in the understanding of the various cellular entry routes and the endocytic process by using the application of FRET in the study of membrane-mediated fusion and factors responsible for membrane binding and fusion dynamics. FRET can differentiate among entry route and endocytic mechanisms, including micelle–membrane interactions (Lee et al. 2013), SNARE-mediated membrane fusion (Diao et al. 2011), M4–64 dye for endocytosis in the plant (Griffing 2008), receptor-mediated (Yang et al. 2006), caveolin and between dependent (McLaughlin et al. 2006) and independent pathways (Glebov et al. 2006).

10.7.1.3 Molecular Trafficking and Localization

FRET a valuable technique for studying trafficking and localization of a cargo (chemo/gene/therapy) delivery through the endocytic pathways. FRET is widely used to monitor the dynamics, location, and quantification of molecules in the living cell (Sekar and Periasamy 2003; Talati et al. 2014). In live-cell imaging, FRET combined with a range of techniques for interrogating the nature of interactions and localization. To investigate diffusion, FCS and FRET are combined techniques for measuring intracellular Gag-Gag interaction and trafficking. It is unknown that in Rous sarcoma virus Gag protein causes retrovirus assembly, whether initiates through interactions between Gag proteins at the plasma membrane or in the cytosol. Cytosolic Gag-Gag interactions were measured by FCS-FRET techniques (Larson et al. 2003). FRET-FLIM used for Imaging and Spatiotemporal localization of molecular interactions between actin-binding protein ezrin and protein kinase C (PKC) α to study the high subcellular resolution of breast carcinoma cells (Peter and Ameer Beg 2004). Combining FRET with atomic force microscopy (AFM) was carried out for high-resolution imaging and localization of viable biological samples (Vickery and Dunn 2001). FRET can also be used for trafficking and localization of individual cellular pathways and to image the spatial and temporal dynamics of protein interactions and intracellular fate of siRNA and to monitor events of the cell cycle, cytokinesis, and cancer progression (Jiang and Zhang 2010).

10.7.2 Measurement of Flux

Dynamic metabolite events and their Quantitative measurement with cellular and subcellular granularity are important to uncover metabolic networks, especially in the context of flux regulation. Metabolic flux may be assessed by recording mass transfer or by using a spectroscopic tracer. Specifically, how does the proximity of endogenous recombinant protein to a target cell monitor the flux and concentrations in a dynamic metabolic network? This understanding is especially important for chimeric recombinant protein to trigger the measurement of flux in real time. The FRET-based biosensors are non-invasively delivered to a subcellular location with spatiotemporal resolution. In order to discover and investigate flux analysis, this biosensor is applied to obtained imaging-based metabolite analysis with genetically encoded nanosensors. This sensor demonstrated the quantitative analysis of cytosolic metabolite levels as well as flux rates, which are composed of the rate of influx minus the rate of metabolism. This genetically encoded tool is best combined with steady-state measurements to obtain a comprehensive characterization of metabolite dynamics. In addition, spectral FRET analysis has been applied to study the Spatiotemporal regulation of various cellular processes. The glucose biosensor has successfully been used for dynamic imaging of glucose flux impedance in Arabidopsis plants for gene discovery (Chaudhuri et al. 2011), and analysis of regulatory networks, for example, by screening mutants (Bermejo et al. 2010). To quantify pyruvate concentration, transport, production, and mitochondrial pyruvate consumption in intact individual cells, genetically encoded FRET-based pyruvate biosensor has been introduced. The pyruvate biosensor demonstrated the mitochondrial flux in single intact cells, allowing the quantification of pyruvate transport, intracellular pyruvate levels, and the rate of glycolytic pyruvate production at a cellular level in real time (San Martin et al. 2014).

10.8 Conclusion

In this chapter, we have discussed the biological implications of several FPs and FRET, which are commonly used as biosensors in imaging. The discovery of FPs has advances the molecular localization and enabled in vivo elucidation of many biological processes. To understand the underlying molecular mechanism of cancer, immunological, neurological diseases, drug discovery, and fluxes, FRET Pairs has played a critical role. Modern biosensors are based on FRET probes and integrated into other techniques that are commonly used in in vivo imaging. Combinations of FPs in FRET pairs and the extension of fluorophore library will facilitate FRET pairs for photobleaching, resistant to degradation over time, negligible toxicity, more easily introduced into the cellular environment, and even more sensitive and robust. FPs-FRET based biosensors are non-invasively introduced into the cell and are excellently suited to an intracellular environment. Because it showed sensitivity and robust means of quantitative and dynamic measurement in cells, researchers have positioned FRET-based genetically encoded nanosensors at the top for intracellular investigation of analytes and dynamic biological processes in vivo in real time. Finally, the use of FPs as a FRET will rule out the concern of toxicity in a live cell. Over the long term, this FRET-based genetically encoded tools can be used in medical research and possibly create innovative tools for clinical application.

References

Adam V, Berardozzi R, Byrdin M, Bourgeois D (2014) Phototransformable fluorescent proteins: future challenges. Curr Opin Chem Biol 20:92–102

Barondeau DP, Putnam CD, Kassmann CJ, Tainer JA, Getzoff ED (2003) Mechanism and energetics of green fluorescent protein chromophore synthesis revealed by trapped intermediate structures. Proc Natl Acad Sci USA 100:12111–12116

Bermejo C, Haerizadeh F, Takanaga H, Chermak D, Frommer WB (2010) Dynamic analysis of cytosolic glucose and ATP levels in yeast using optical sensors. Biochem J 432(2):399–406

Bhuckory S, Mattera L, Wegner KD, Qiu X, Wu YT, Charbonniere LJ, Hildebrandt N (2016) Direct conjugation of antibodies to the ZnS shell of quantum dots for FRET immunoassays with low picomolar detection limits. Chem Commun 52:14423–14425

Carpentier P, Violot S, Blanchoin L, Bourgeois D (2009) Structural basis for the phototoxicity of the fluorescent protein killer red. FEBS Lett 583:2839–2842

Chaudhuri B, Hörmann F, Frommer WB (2011) Dynamic imaging of glucose flux impedance using FRET sensors in wild-type Arabidopsis plants. J Exp Bot 62(7):2411–2417

Chu J, Haynes RD, Corbel SY, Li P, Gonzalez-Gonzalez E, Burg JS, Davidson MW (2014) Non-invasive intravital imaging of cellular differentiation with a bright red-excitable fluorescent protein. Nat Methods 11:572–578

Clegg RM (2009) Forster resonance energy transfer-FRET what is it, why do it, and how it's done. Lab Tech in BMB 33:1–57

Delagrave S, Hawtin RE, Silva CM, Yang MM, Youvan DC (1995) Red-shifted excitation mutants of the green fluorescent protein. Biotechnology 13:151–154

Demarco IA, Periasamy A, Booker CF, Day RN (2006) Monitoring dynamic protein interactions with photoquenching FRET. Nat Methods 3:519–524

Diao J, Ishitsuka Y, Bae WR (2011) Single-molecule FRET study of SNARE-mediated membrane fusion. Biosci Rep 31(6):457–463

Frommer WB, Davidson MW, Campbell RE (2009) Genetically encoded biosensors based on engineered fluorescent proteins. Chem Soc Rev 38:2833–2841

Giordano L, Jovin TM, Irie M, Jares-Erijman EA (2002) Diheteroarylethenes as thermally stable photoswitchable acceptors in photochromic fluorescence resonance energy transfer (pc FRET). J Am Chem Soc 124:7481–7489

Glebov OO, Bright NA, Nichols BJ (2006) Flotillin-1 defines a clathrin-independent endocytic pathway in mammalian cells. Nat Cell Biol 8:46–54

Griffing LR (2008) FRET analysis of transmembrane flipping of FM4–64 in plant cells: is FM4–64 a robust marker for endocytosis? J Microsc 231:291–298

Gross LA, Baird GS, Hoffman RC, Baldridge KK, Tsien RY (2000) The structure of the chromophore within DsRed, a red fluorescent protein from coral. Proc Natl Acad Sci U S A 97:11990–11995

Haddock SH, Moline MA, Case JF (2010) Bioluminescence in the sea. Annu Rev Mar Sci 2:443–493

Heim R, Prasher DC, Tsien RY (1994) Wavelength mutations and posttranslational autoxidation of green fluorescent protein. Proc Natl Acad Sci U S A 91:12501–12504

Heim R, Tsien RY (1996) Engineering green fluorescent protein for improved brightness, longer wavelengths, and fluorescence resonance energy transfer. Curr Biol 6:178–182

Hotzer B, Medintz IL, Hildebrandt N (2012) Fluorescence in nanobiotechnology: sophisticated fluorophores for novel applications. Small 8:2297–2326

Jares-Erijman EA, Jovin TM (2003) FRET imaging. Nat Biotechnol 21:1387–1395

Jiang S, Zhang Y (2010) Upconversion nanoparticle-based FRET system for the study of si RNA in live cells. Langmuir 26:6689–6694

Karasawa S, Araki T, Nagai T, Mizuno H, Miyawaki A (2004) Cyanemitting and orange-emitting fluorescent proteins as a donor/acceptor pair for fluorescence resonance energy transfer. Biochem J 381:307–312

Kremers GJ, Goedhart J, Van Munster EB, Gadella TW (2006) Cyan and yellow super fluorescent proteins with improved brightness, protein folding, and FRET Forster radius. Biochemistry 45:6570–6580

Krutzik PO, Nolan GP (2003) Intracellular phospho-protein staining techniques for flow cytometry: monitoring single cell signaling events. Cytometry A 55:61–70

Lam AJ, St-Pierre F, Gong Y, Marshall JD, Cranfill PJ, Baird MA, McKeown MR, Wiedenmann J, Davidson MW, Schnitzer MJ (2012) Improving FRET dynamic range with bright green and red fluorescent proteins. Nat Methods 9:1005–1012

Larson DR, Ma YM, Vogt VM, Webb WW (2003) Direct measurement of gag-gag interaction during retrovirus assembly with FRET and fluorescence correlation spectroscopy. J Cell Biol 162:1233–1244

Lee SY, Tyler JY, Kim S, Park K, Cheng JX (2013) FRET imaging reveals different cellular entry routes of self-assembled and disulfide-bonded polymeric micelles. Mol Pharm 10:3497–3506

Li-En Jao D, Yu Chen K (2002) Subcellular localization of the hypusine-containing eukaryotic initiation factor 5A by immunofluorescent staining and green fluorescent protein tagging. J Cell Biochem 86:590–600

Masharina A, Reymond L, Maurel D, Umezawa K, Johnsson KA (2012) Fuorescent sensor for gaba and synthetic gaba (b) receptor ligands. J Am Chem Soc 134:19026–19034

McLaughlin NJD, Banerjee A, Kelher MR, Gamboni-Robertson F, Hamiel C, Sheppard FR, Moore EE, Silliman CC (2006) Platelet-activating factor-induced clathrin-mediated endocytosis requires beta-arrestin-1 recruitment and activation of the p 38 MAPK signalosome actin bundle formation. J Immunol 176:7039–7050

Miyawaki A (2003) Visualization of the spatial and temporal dynamics of intracellular signaling. Dev Cell 4:295–305

Miyawaki A (2011) Development of probes for cellular functions using fluorescent proteins and fluorescence resonance energy transfer. Annu Rev Biochem 80:357–373

Miyawaki A, Llopis J, Heim R, McCaffery JM, Adams JA, Ikura M, Tsien RY (1997) Fluorescent indicators for Ca2+ based on green fluorescent proteins and calmodulin. Nature 388:882–887

Miyawaki A, Nagai T, Mizuno H (2003) Mechanisms of protein fluorophore formation and engineering. Curr Opin Chem Biol 7:557–562

Murakoshi H, Shibata AC, Nakahata Y, Nabekura J (2015) A dark green fluorescent protein as an acceptor for measurement of Forster resonance energy transfer. Sci Rep 5:15334

Niino Y, Hotta K, Oka K (2009) Simultaneous live-cell imaging using dual FRET sensors with single excitation light. PLoS One 4:e6036

Okamoto KI, Nagai T, Miyawaki A, Hayashi Y (2004) Rapid and persistent modulation of actin dynamics regulates postsynaptic reorganization underlying bidirectional plasticity. Nat Neurosci 7:1104–1112

Okumoto S, Versaw W (2017) Genetically encoded sensors for monitoring the transport and concentration of nitrogen-containing and phosphorus-containing molecules in plants. Curr Opin Plant Biol 39:129–135

Onuki R, Nagasaki A, Kawasaki H, Baba T, Uyeda TQ, Taira K (2002) Confirmation by FRET in individual living cells of the absence of significant amyloid β-mediated caspase 8 activation. Proc Natl AcadSci USA 99:14716–14721

Peter M, Ameer Beg SM (2004) Imaging molecular interactions by multiphoton FLIM. Biol Cell 96:231–236

Phillip Y, Kiss V, Schreiber G (2012) Protein-binding dynamics imaged in a living cell. Proc Natl Acad Sci U S A 109:1461–1466

Piston DW, Kremers GJ (2007) Fluorescent protein FRET: the good, the bad and the ugly. Trends Biochem Sci 32:407–414

Poole E, Strappe P, Mok HP, Hicks R, Lever AML (2005) HIV-1 gag–RNA interaction occurs at a perinuclear/centrosomal site, analysis by confocal microscopy and FRET. Traffic 6:741–755

Prasher DC, Eckenrode VK, Ward WW, Prendergast FG, Cormier MJ (1992) Primary structure of the Aequorea victoria green-fluorescent protein. Gene 111:229–233

San Martin A, Ceballo S, Baeza-Lehnert F, Lerchundi R, Valdebenito R, Contreras-Baeza Y, Barros LF (2014) Imaging mitochondrial flux in single cells with a FRET sensor for pyruvate. PLoS One 9:e85780

Sekar RB, Periasamy A (2003) Fluorescence resonance energy transfer (FRET) microscopy imaging of live cell protein localizations. J Cell Biol 160:629–633

Shaner NC, Campbell RE, Steinbach PA, Giepmans BN, Palmer AE, Tsien RY (2004) Improved monomeric red, orange, and yellow fluorescent proteins derived from Discosoma sp. red fluorescent protein. Nat Biotechnol 22:1567–1572

Shcherbakova DM, Hink MA, Joosen L, Gadella TW, Verkhusha VV (2012) An orange fluorescent protein with a large stokes shift for single-excitation multicolor FCCS and FRET imaging. J Am Chem Soc 134:7913–7923

Shcherbo D, Souslova EA, Goedhart J, Chepurnykh TV, Gaintzeva A, Shemiakina I, Chudakov DM (2009) Practical and reliable FRET/FLIM pair of fluorescent proteins. BMC Biotechnol 9:24

Shi L, De Paoli V, Rosenzweig N, Rosenzweig (2006) Synthesis and application of quantum dots FRET-based protease sensors. J Am Chem Soc 128:10378–10379

Shimomura O, Johnson FH, Saiga Y (1962) Extraction, purification and properties of aequorin, a bioluminescent protein from the luminous hydromedusan, Aequorea. J Cell Comp Physiol 59:223–239

Specht EA, Braselmann E, Palmer AE (2017) A critical and comparative review of fluorescent tools for live-cell imaging. Annu Rev Physiol 79:93–117

Stryer L (1978) Fluorescence energy transfer as a spectroscopic ruler. Annu Rev Biochem 47:819–846

Subach FV, Zhang L, Gadella TW, Gurskaya NG, Lukyanov KA, Verkhusha V (2010) Red fluorescent protein with reversibly photoswitchable absorbance for photochromic FRET. Nat Chem Biol 17:745–755

Talati R, Vanderpoel A, Eladdadi A, Anderson K, Abe K, Barroso M (2014) Automated selection of regions of interest for intensity-based FRET analysis of transferrin endocytic trafficking in normal vs. cancer cells. Methods 66:139–152

Tsien RY (1998) The green fluorescent protein. Annu Rev Biochem 67:509–544

Ueda Y, Kwok S, Hayashi Y (2013) Application of FRET probes in the analysis of neuronal plasticity. Front Neural Circ 7:163

Vickery SA, Dunn RC (2001) Combining AFM and FRET for high-resolution fluorescence microscopy. J Microsc 202:408–412

Wall MA, Socolich M, Ranganathan R (2000) The structural basis for red fluorescence in the tetrameric GFP homolog DsRed. NatStruct Mol Biol 7:1133–1138

Ward WW, Chalfie M, Kain SR (2005) Biochemical and physical properties of green fluorescent protein. Green fluorescent protein: properties, applications, and protocols. Wiley Hoboken USA 2:39–65

Yang J, Chen H, Vlahov IR, Cheng JX, Low PS (2006) Evaluation of disulfide reduction during receptor-mediated endocytosis by using FRET imaging. Proc Natl Acad Sci U S A 103:13872–13877

Zhang J, Campbell RE, Ting AY, Tsien RY (2002) Creating new fluorescent probes for cell biology. Nat Rev Mol Cell Biol 3:906–918

Zhao Y, Yang Y (2015) Profiling metabolic states with genetically encoded fluorescent biosensors for NADH. Curr Opin Biotechnol 31:86–92

Probes for Detection of Free Radicals 11

Neha Agrawal and Mohd. Mohsin

Abstract

Reactive species are reactive molecules, having one or two lone pairs that are naturally generated in biological environment via endogenous and exogenous method. These are non-avoidable by-product, highly reactive which are mainly produced by the Electron Transport Chain (ETC) of mitochondria by aerobic respiration. In this chapter, we explore the multi-faceted pathological roles of ROS in the living cell. These reactive species have bimodal action in both physiological and pathological processes. In regard to the good side of ROS, it plays a significant role in signaling, immune response, homeostasis, cell growth, signal transduction, and other physiological responses. Along with the good side, it has also the potential bad side, abnormal concentrations of these reactive molecules lead to the interference in redox homeostasis, which can induce oxidative stress, damaging intracellular components, and alter protein expression. The ugly effects of these mitochondrial reactive species can lead to devastating cardiomyopathies, pulmonary disorder, and neurodegenerative diseases. Various approaches have been developed so far for the detection of ROS but due to their short life span and low availability in biological system, these approaches are not very specific. In recent study, many fluorescent and mitochondrial targeting probes came into existence and these probes have the potential ability to quantified these free radicals very rapidly and efficiently. Furthermore, in this chapter, we focus on various methods like synthetic sensors, florescent probes, chemiluminescent probes, fiber optic sensors, electrochemical probes, and various alternative approaches that are used to analyze and quantify free radicals in vivo.

N. Agrawal · M. Mohsin (✉)
Department of Biosciences, Jamia Millia Islamia, New Delhi, India
e-mail: mmohsin1@jmi.ac.in

M. Mohsin et al. (eds.), *Nanobiosensors for Agricultural, Medical and Environmental Applications*, https://doi.org/10.1007/978-981-15-8346-9_11

Keywords

Free radicals · FRET · Fluorescent probes · Chemiluminescent probes · Chemical reporters

11.1 Introduction

Free radicals can be atoms, ions, and molecules that contain one or more unpaired valence electron in the outer orbit. This unpaired electron makes them highly unstable and extremely reactive. The electron comes from the oxidation process of fats, lipid, cell membrane, nucleic acid, and different biological pathways *via* altering the fundamental structure of affected molecules. These free radicals mainly cause oxidative damage and scavenging effect in cells. There are many types of free radicals like Reactive Oxygen Species (ROS) and Reactive Nitrogen Species (RNS). These reactive species are generated through the normal biological process via mitochondrial respiration and some of these having a biological importance like signal transduction, neurotransmission, smooth muscle relaxation, peristalsis, platelet aggregation, blood pressure modulation, immune system control, learning and memory, energy production, phagocytosis, cell signaling, synthesis of various important biological compound and maintain homeostasis in cellular level (Gielis et al. 2011). Overproduction of these reactive species have a destructive impact on the cellular level and lead to aging, cellular damage, protein dysfunction, apoptosis, mitochondrial oxidative stress, and DNA damage (Fig. 11.1).

Due to the potential imperative role of these free oxidative radicals in redox homeostasis, ranging from their role in the signaling process, modulation of cell proliferation to the more unexpected role in the treatment of certain cancers; the measurement and detection of these reactive species are more important nowadays. Here, we focus on diverse detection methods of these reactive species. The major hindrance in their detection is diminutive life spans which ranging from nanosecond to millisecond and low availability in the biological system (Table 11.1).

11.2 Generation of ROS and RNS

Reactive oxygen species (ROS) are chemically formed non-radical derivatives of oxygen (O_2) that are more reactive than molecular oxygen. Reactive oxygen species (ROS; e.g., $O_2^{\bullet-}$, H_2O_2, $OH^{\bullet}$, $^1O_2$) are partially reduced or excited forms of atmospheric oxygen. A primary ROS is a superoxide (O^{2-}), which is produced by the reduction of one-electron of molecular oxygen (Singh et al. 1992). Hydrogen peroxide (H_2O_2) is created by the reduction of superoxide via superoxide dismutase enzyme(MnSOD) by simultaneous oxidation and reduction process. Hydroxyl radicals are also generated by the reduction of H_2O_2 by the Fenton reaction as

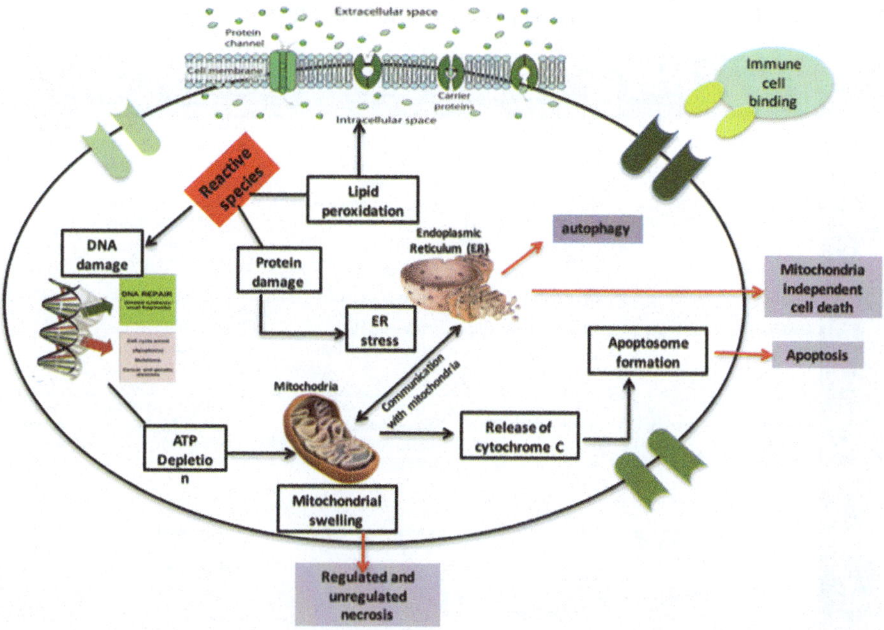

Fig. 11.1 Lethal consequences of Reactive species in a cell

well as hypochlorous acid with the help of myeloperoxidase enzyme Arnhold 2004). Mitogenic signalling begins at the cell surface with the ligand-dependent activation of receptor tyrosine kinases. These tyrosine kinase receptors help in activating MAP kinase cascades which have a crucial role in cell proliferation. These cascades lead to the production of H_2O_2 from several enzyme catalysts, including the NADPH oxidases (Park et al. 2006). ROS are highly reactive molecules that originate mainly from the mitochondrial electron transport chain and endoplasmic reticulum via many enzymatic pathways. ROS and oxygen radicle are formed mainly by exogenous (heat, trauma, UV light, ozone, smog, aging, and therapeutic drugs) and endogenous (the by-product of metabolic pathways and functional generation by host defense cells) methods (Fig. 11.2).

Reactive nitrogen species (RNS) like nitric oxide ($^{\cdot}NO$) nitrogen dioxide radical ($^{\cdot}NO_2$), but also non-radicals, like peroxynitrite anion (ONOO), peroxynitrous acid (ONOOH), nitrosoperoxycarbonate anion ($ONOOCO_2$), nitronium cation (NO_2^+), and dinitrogen trioxide (N_2O_3) are continuously generated in small quantities on normal cellular processes. $NO^{\bullet}$ plays a significant role in cellular signaling, vasodilation, and immune response (White and Marletta 1992). It is a highly reactive small-uncharged molecule containing one unpaired electron. RNS can readily diffuse across the membrane due to its uncharged state. Endogenous $NO^{\bullet}$ is formed in the biological tissues via the action of NOS where L-arginine and oxygen are converted into $NO^{\bullet}$ and citrulline via a five-electron oxidative process. A main source of RNS

Table 11.1 Reactivity and scavenging properties of ROS and RNS

		Structure	$t_{1/2}$	Migration distance	Production site	Disease
REACTIVE OXYGEN SPECIES	Superoxide		~1-4μs	300Å	Electron transport chain, Mitochondria, chloroplast, Apoplast, peroxisomes	cause an immunodeficiency syndrome called chronic granulomatous disease, lung diseases such as Acute Respiratory Distress Syndrome (ARDS) or Chronic obstructive pulmonary disease (COPD)
	Hydroxyl radical		~1ns	10 Å	Iron and H_2O_2 (Fenton reaction)	Increasing the risk of cardiovascular disease.
	Hydrogen peroxide		~>1ms	>10,000 Å	Mitochondria, peroxisomes, chloroplast, Apoplast, cytosol	Aging, Alzheimer disease, cardiovascular disorder, Ischemia/reperfusion injury
	Singlet oxygen		~1-4 μs	300Å	Membrane, Chloroplast, nuclei	Heme oxidation, Oxidation of iron-sulfur centers in proteins
REACTIVE NITROGEN SPECIES	Peroxynitrate	OONO⁻	~1ms	_	Macrophages, mitochondria	DNA damage, disruption of membrane-associated signaling via alter amino acids residues
	Nitric oxide	$NO^{\bullet}$	~1 sec	_	Phagocytes, lysosomes	Thiol oxidation, carcinogenic via altering gene expression
	Dinitrogen trioxide	N_2O_3	~1 μs	_	Peroxysomes, Mitochondria	Lipid peroxidation which leads to the formation of protein adducts and DNA damage
	Nitrosoperoxy carbonate anion	ONOOCO 2	-		Mitochondria, peroxysomes	Disruption of membrane, depletion of ATP and NAD(H) causes cell death
	Nitronium cation	NO_2^+	-		lysosomes	Damage various ion transport system

synthesis in a biological system is $NO^{\bullet}$ that can reduce both microbial DNA replication and cellular respiration. Overproduction of RNS causes nitrosative stresses, which include mitochondrial dysfunction, cell death or injury, and altered structure and function of proteins. Peroxynitrite is furthermost reactive and potentially harmful RNS, and it has influential oxidizing and nitrating actions. These anions target DNA, protein, lipids, thiols, and amino acids (Suzuki et al. 1997) (Fig. 11.3).

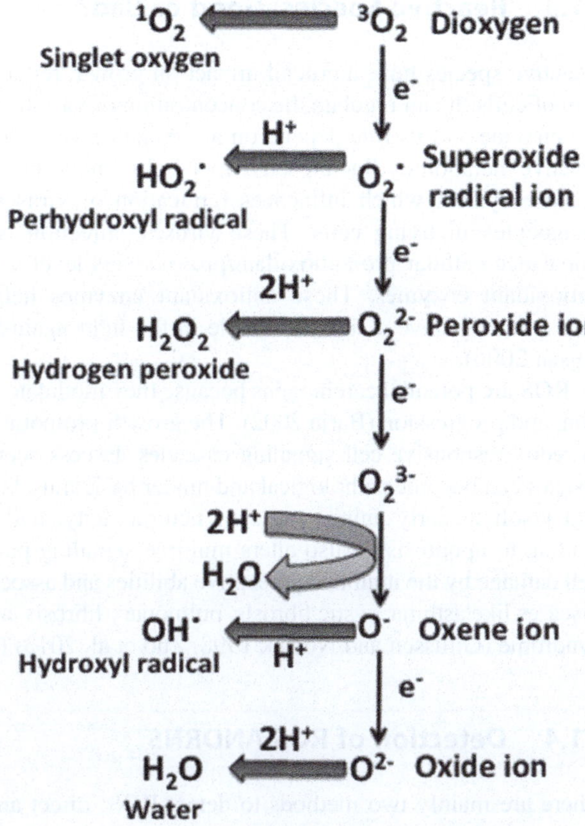

Fig. 11.2 Generation of different ROS by energy transfer from ground state triplet oxygen to oxide ion

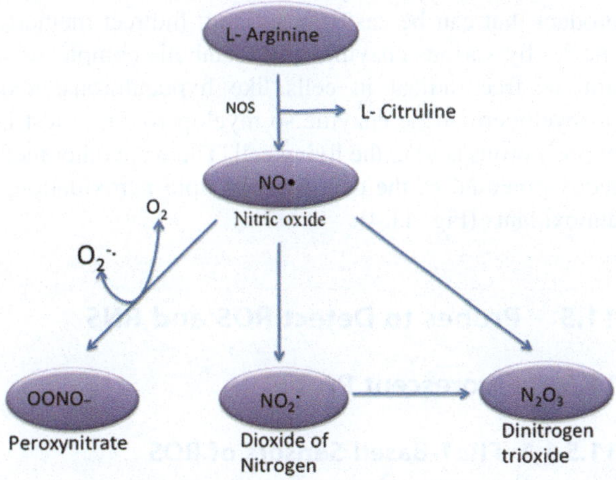

Fig. 11.3 Production of reactive nitrogen species in cell

11.3 Reactive Species: Good or Bad?

Reactive species have a crucial impact on growth regulation and metabolic regulation of cells. It can regulate the concentration of various hormones and activation of the enzyme and its role depend on an oxidant form and concentration of particular reactive metabolites. Such reactive metabolites have an important role in the immune defense system which influences replication of viruses and induce activation of phagocytes in living cells. These viruses' infection can affect the host cell by imbalance cellular pro-antioxidant/pro-oxidants level and inhibit the production of antioxidant enzymes. These antioxidant enzymes help in dealing with foreign invasion may have a significant role in the fight against viral diseases (Waris and Ahsan 2006).

ROS are potential carcinogens because they modulate mutagenesis, tumor promotion, and progression (Barja 2002). The growth-promoting effects of ROS are related to redox-responsive cell signaling cascades. Excess accumulation of these reactive species can become pathological and hinder basic muscle function, multiple sclerosis and result in early aging, cause of neurotoxicity, and DNA Damage, ultimately leading to apoptosis. It also alters multiple signaling pathways that result in neural cell damage by the inhibited perceptive abilities and associated with many pulmonary diseases like asthma, cystic fibrosis, pulmonary fibrosis, and adult pulmonary distress syndrome (Gillissen and Nowak 1998, Zuo et al. 2013) (Table 11.2).

11.4 Detection of ROS ANDRNS

There are mainly two methods to detect ROS; direct and indirect methods. Direct assays include chemiluminescence, nitro blue tetrazolium(NBT) test, flow cytometry, Electron Spin Resonance (ESR), and xylenol orange-based assay. In direct approaches, these probes react with cellular reactive species and form a stable product that can be easily quantified. Indirect methods include analyzing reactive species by various enzymes and metabolic complexes which involve in the generation of free radical in cells like hypochlorous acid is formed with the help of myeloperoxidase enzyme so myeloperoxidase test is used to detect the level of hypochlorous acid in the living cell. There are other methods for measurement of the redox potential of the cell, level of lipid peroxidation, and levels of chemokine's, antioxidants (Fig. 11.4).

11.5 Probes to Detect ROS and RNS

11.5.1 Fluorescent Probe

11.5.1.1 FRET-Based Sensors of ROS

Förster Resonance Energy Transfer or Fluorescence Resonance Energy Transfer (FRET) is a phenomenon which describes the transmission of energy between two light-sensitive fluorophores of a different color. FRET is not only restricting to

Table 11.2 Harmful and beneficial role of ROS and RNS

	Biological and physiological activity	Effect of ROS
ROS as good	Immune System	Activates toll like receptor-4 or other immune-related receptors and fight against foreign entity
	Cellular Activities	Involved in cellular response to stress or regulates mitochondrial function, expression and transcriptions of certain stress proteins and antioxidant levels. Required for proper cell growth, cell proliferation as well as programed cell death
ROS as bad	Cancer	Induces DNA damage and regulates HIF-1a level which involved in cell apoptosis and autoimmune disorder
	Pulmonary Diseases	Enhances inflammation response and alter the normal cellular process, damages diaphragm function Contributes to pulmonary diseases and vascular pathologies
	DNA Damage	Induces mutagenesis and causes the protein dysfunction
	Neurodegenerative Disease	Stimulate various enzymatic pathways like myeloperoxidases which is associated with Parkinson's disease, Alzheimer's disease and multiple sclerosis
	Cardiovascular Diseases	Involved in cell signaling and transduction, causing hypertension, oxidative stress

fluorescence but it can also occur in phosphorescence as well. Fluorescence is a light emission phenomenon by fluorophore molecules when they absorb photons or electromagnetic radiation. The light absorbed by fluorophore molecules mostly has a shorter wavelength than emitted light, because the conversion of excitation energy has taken place between two fluorophore pair in non-radioactive manner. When fluorophore absorbs light, it may lead to the excitation of electron, and this excited state of the electron is extremely unstable. The electron relaxes back to its ground

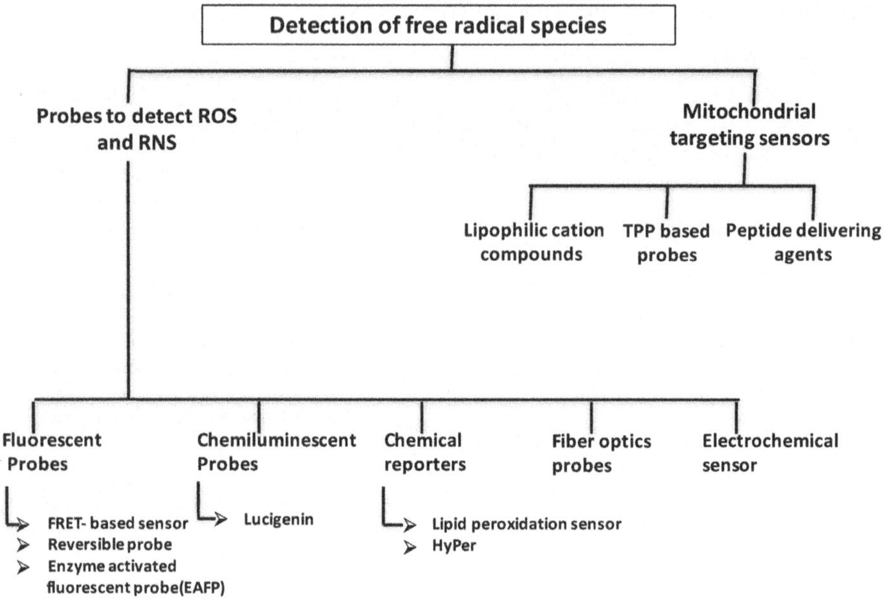

Fig. 11.4 Various methods for detection of ROS/RNS

state from the excited state within a fraction of second. In this process, some energy has lost in these levels due to vibrational relaxation. This effect was first demonstrated by Theodor Förster in 1948 (Förster 1948) and hence termed Förster Resonance Energy Transfer (FRET).

FRET is also accounted for the loss of energy when excitation energy with lower wavelength are compared with higher absorbing wavelength. There is an exception to this rule, when two or more photons exploit with high intensity on a fluorophore, it may lead to the excitation of two or more photons. When two chromophores have come in close proximity range (1 nm–10 nm), some part of the energy can be transferred by non-radiation manner from donor fluorophore to acceptor molecules through direct dipole interaction. The first basic fundamental for FRET to occur is an overlapping spectrum of the emission wavelength of donor molecules, and the excitation wavelength of acceptor molecules. This overlapping spectrum is required for the appropriate energy transfer between the two compounds. For the detection of FRET, it is usually assumed that 30% or more overlap is sufficient to give reliable results (Lovell et al. 2009). Along with the sufficient proximity between the two chromophores (1 nm to10 nm), a correct orientation between the dipole of the chromophores is necessary for a high FRET ratio. The orientation of the dipoles must be perpendicular to each other for maximum energy transfer (Fig. 11.5). Protein interactions and many other biological processes take place within the spatial range of FRET. So, it allows monitoring and visualizing various biological molecules and their interactions with in a cell in a non-invasive manner.

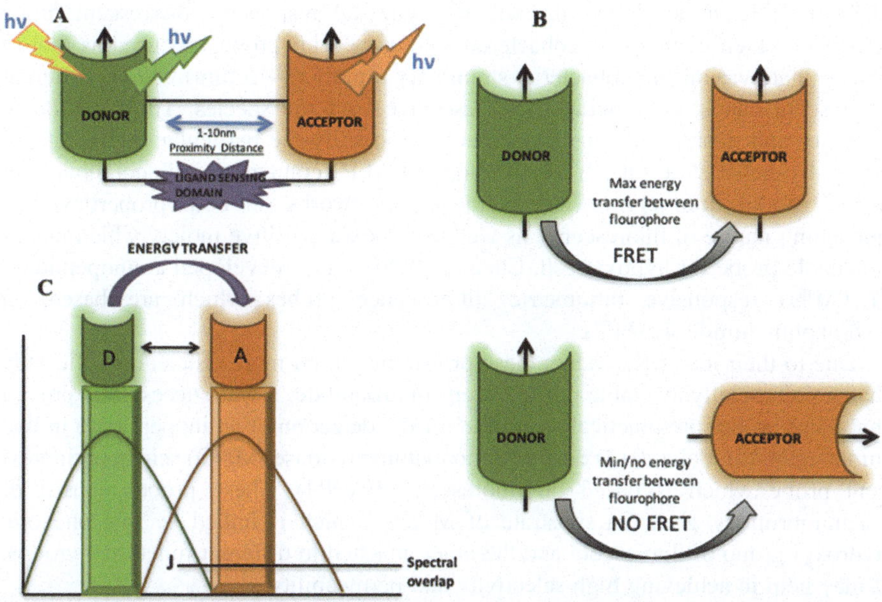

Fig. 11.5 (**a**) Schematic representation of FRET where donor transfers its energy to the nearby acceptor, (**b**) orientation of two fluorophores, (**c**) Donor–Acceptor overlapping spectrum

11.5.1.2 Reversible Probe

Reversible probes are used for monitoring the pathophysiological process of free radicals *in vitro* as they bind to mitochondrial respiration complex and regulate its function in multiple cycles. Miller et al. (2007) has reported a reversible fluorescent probes (RP1) which help in analyzing reversible redox cycles in vivo. These probes are responsible for a reversible response to multiple oxidation/reduction events, >50-fold fluorescence dynamic range, and excitation/emission profiles in the visible region to minimize cellular damage and autofluorescence. Koide et al. (2012), developed a reversible near-infrared fluorescence probe (RP2) for free radical which is based on bio reporter tellurium (Te). RP2 is oxidized in the presence of various ROS, but this oxidized form is quickly reduced by glutathione to regenerate RP2. This redox-induced reversible NIR-fluorescence response of RP2 was allowed the detection of free radicals which generate by the endogenous process in living cells.

11.5.1.3 Enzyme Activated Fluorescent Probes

Enzyme Activated Fluorogenic Probes (EAFP) provide the prospect for monitoring the response of various cancer therapy in real time. The strategy involved in the enzyme sensing process, samples being examined are treated with EAFPs that react with target enzymes. The analysis is done by imaging, SDS-PAGE, etc. or purification of the target using tags attached to EAFPs. They can offer direct biochemical confirmation about the enzymes which involve in ROS production. Beatty et al., in

2013 have been developed the sulfatase enzyme responsive fluorescent probe (EAPF1) which contains mycobacterial sulfatases. This probe was used as tool to discriminate various mycobacterial strain very rapidly (5–15 min). On gel different fluorescent bands are generated by these mycobacterial species. The variation in intensity and position of these bands help to identify the particular mycobacterial strain. Komatsu et al. (2010) has developed EAFP2 containing indolequinone unit and a rhodol fluorophore. The indolequinone works on both properties, i.e., quenching nature of fluorescence as well as hypoxia-sensitive moiety which makes it suitable probe for hypoxia cell. Chen et al. 2013 have developed aminopeptidase N (APN) responsive ratiometric fluorescence probes which are based on 1, 8-naphthalimide (EAFP3).

Due to their lower K_m value, these probes may bind more efficiently with APN than commercially available probe L-Leu-p-nitroanilide. It was successfully proved that these probes are practically feasible for the detection of aminopeptidase in live tumor cells. Li et al. (2014) designed monoamine oxidase (MAO) sensitive fluorescent probe which is based on fluorescein (EAFP4). These probes consist of 3-aminopropoxy group, a substrate of MAOs, which is linked by one phenolic hydroxyl group of fluorescein and the other attached to different functional groups. It may help in achieving high selectivity and permeability.

11.6 Fluorescent Probes Developed So Far

11.6.1 Hydroethidine

It is reported that Hydroethidine (dihydroethidium; DHE) has high relative specificity for ROS. So, it used as a fluorescent probe for detecting $O_2^{\bullet-}$ ions. It is commonly used in analyze of the oxidative burst in phagocytes (Fig. 11.6). It has the ability to cross the mitochondrial membranes. $O_2^{\bullet-}$ ions are the initially produced in mitochondria as ROS so HE is oxidized by $O_2^{\bullet-}$ and produce a fluorescent compound, ethidium(E^+) which itself mixes with DNA and its fluorescence increase (Gomes et al. 2005). Hydroethidine (HEt) and its mitochondria-targeted variant

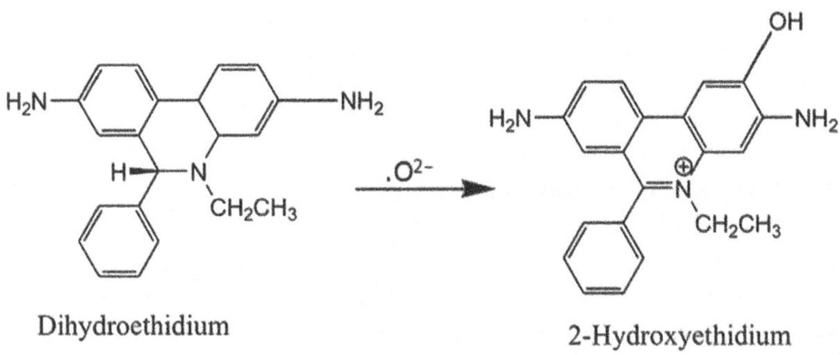

Fig. 11.6 Conversion of non-fluorescent Dihydroethidium to fluorescent 2 Hydroxyl ethidium

Mito-HEt (MitoSOX-red) are frequently used as a chemical reporter for the detection of superoxide (Robinson et al. 2006). Both HEt and Mito-HEt can easily diffuse through lipid bilayers membrane of mitochondria. The initial oxidation of non-fluorescent HEt and Mito-HEt occur in the following steps:

- Production of (HE•+) radical
- Oxidative conversion of this radical in to two positively charged fluorescent by-products, i.e., ethidium (Et+) and 2-hydroxyethidium (2-HO-Et+) (Zhao et al. 2013)

11.6.2 2′, 7′Dichlorfluorescein-Diacetate (DCFH-DA)

2′,7′Dichlorfluorescein-Diacetate (DCFH-DA) has simple and direct chemistry with reactive species mainly H_2O_2.So, this has been commonly used as chemical fluorescent detector for measuring the level of intracellularH_2O_2 (Sundaresan et al. 1995; Frank et al. 2000). 2′, 7′-dichloro-fluorescein (DCF) has been produced by 2′, 7′-dichlorodihydro-fluorescein (H_2DCF) through the oxidation process which is widely used to measure levels of ROS in the intracellular environment. (Fig. 11.7). These compounds are transferred into cells or tissues in the reduced form of a membrane-permeant diacetate (DA) ester. DCF is prone to passive leakage across the membrane and has low specificity. Therefore, a DCF variant with chloromethyl groups (CMH_2DCF) has been developed to reduced passive dye leakage (Koopman et al. 2006). By attaching esterase-cleavable DA groups to CM-H_2DCF its membrane-permeability was improved. DCFH has been shown to react with agents other than H_2O_2 making it more of a measure of total redox status rather than intracellular H_2O_2 (Ischiropoulos et al. 1999; Royall and Ischiropoulos 1993).

11.6.3 Scopoletin (7-Hydroxy-6-Methoxy-Coumarin)

Gajovic-Eichelmann et al. (2003) established an electro monomer (scopoletin) which is a derivative of naturally occurring coumarin (Fig. 11.8). This monomer has some advantages such as good solubility in aqueous solution, no pre-purification or the deoxygenation of the monomer solution, and hydrophilicity of the deposited polymer.

11.6.4 N-Acetyl-3, 7-Dihydroxyphenoxazine (Amplex Red)

It is a non-fluorescent reagent, which undergoes an oxidation reaction in the presence of horseradish peroxidase (HRP) by hydrogen peroxide. Amplex red react with H_2O_2in stoichiometric amount (1:1) and produce fluorescent oxidation product,

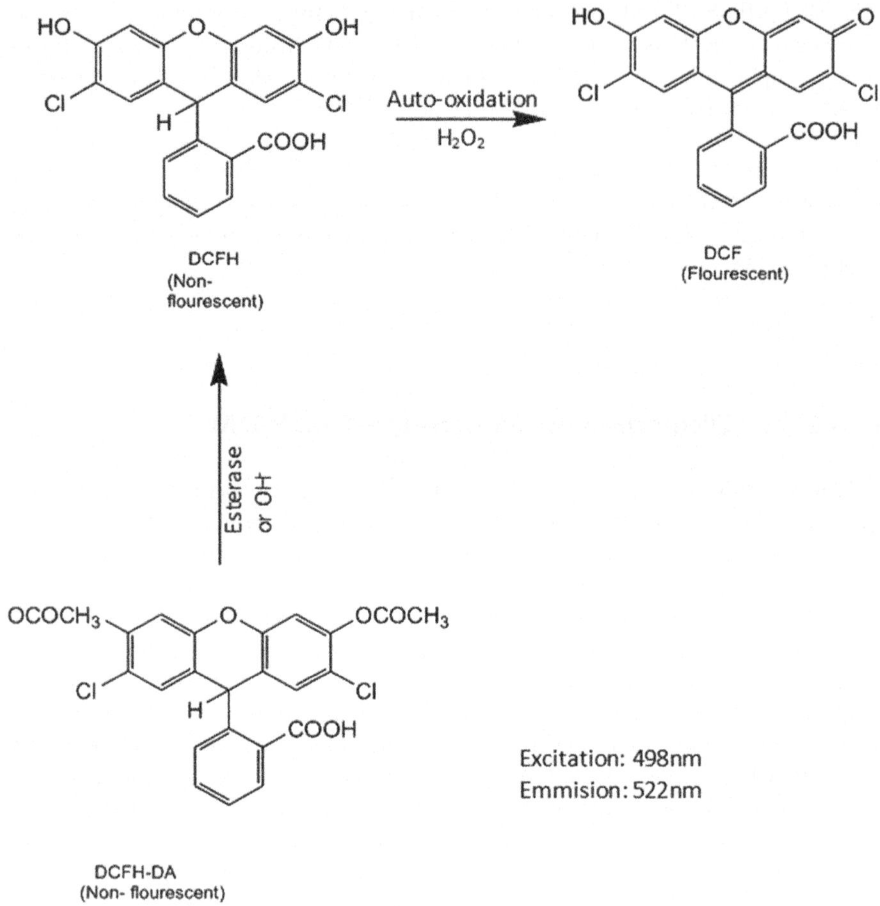

Fig. 11.7 Conversion of 2′, 7′dichlorfluorescein-diacetate into fluorescent 2′, 7′-dichlorofluorescein

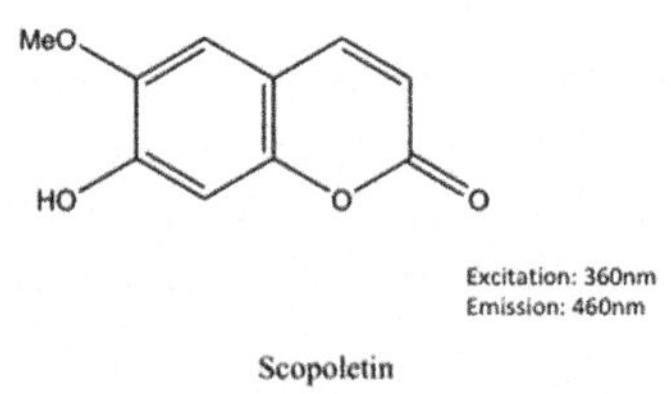

Fig. 11.8 Structure of scopoletin

resorufin (Fig. 11.9). Amplex Red reagent is sensitive and specific for hydrogen peroxide so it is used as a tool for the detection of lower level H_2O_2. It not only detect the level of H_2O_2inactivated phagocytic cells but also in other types of cells like

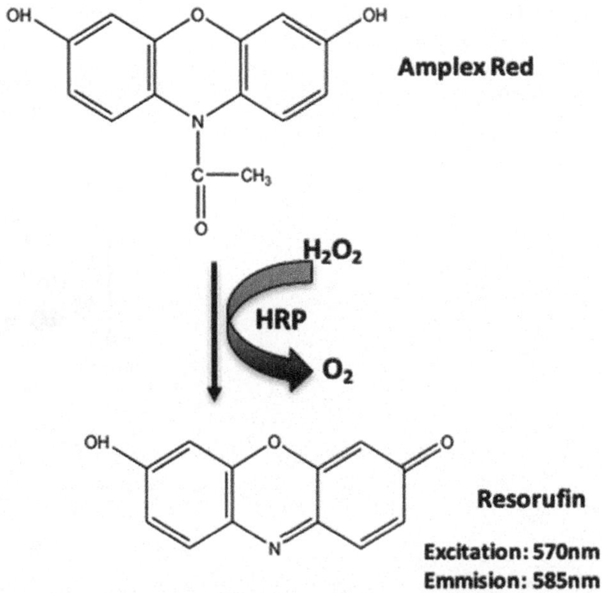

Fig. 11.9 Conversion of non-fluorescent Amplex red into fluorescent Resorufin

activated human leucocytes or even in non-cellular systems (Mohanty et al. 1997; Zhou et al. 1997).

11.6.5 Homovanillic Acid (4-Hydroxy-3-Methoxy-Phenylacetic Acid; HVA)

HVA is also non-fluorescent compound, which is oxidized by H_2O_2in the presence of HRP and convert into a fluorescent dimer (Fig. 11.10). It shows more specificity and sensitivity towards H_2O_2 than scopoletin. This may also be used for the fluorimetric detection of peroxidase activity in various biochemical and biological assays.

11.7 Chemiluminescent Probes

The basic principle of chemiluminescence is the measurement of emitted light produced by a chemical reaction which occurs between chemical reagents and the free oxidant radical generated in the intracellular environment by aerobic respiration. The reaction causes the emission of light, which is directly proportional to the level of free radical. These sensors are more sensitive and used for quantitative analysis of free radicals.

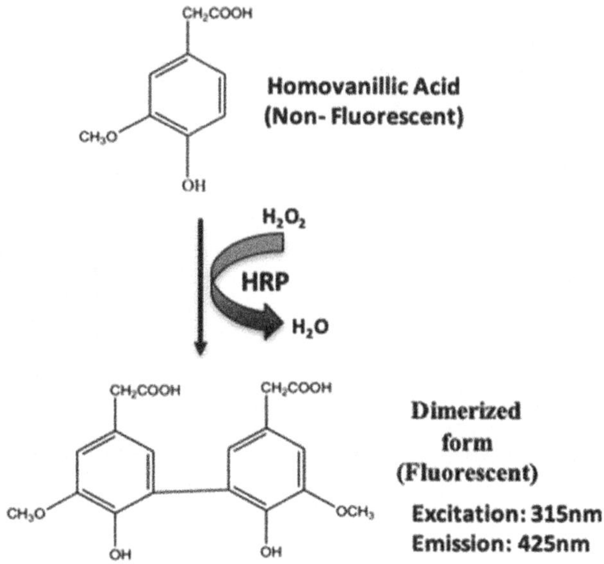

Fig. 11.10 Conversion of non-fluorescent Homovanillic Acid into fluorescent dimer form

11.7.1 Lucigenin (Bis-*N*-Methyl Acridinium Nitrate)

It is an acridine-based lucigenin which is most widely used as a chemiluminescent probe to detect superoxide ions in cell at neutral pH (Allen 1986; Gyllenhammar 1987). The reaction between superoxide and lucigenin generates chemiluminescence in enzymatic and cellular systems. Lucigenin can detect extracellular free radical in various types of cells using reductase enzymes which are involved in aerobic respiration like NADPH oxidase in phagocytic cells. The ETC complex and other various enzymes like NADH-dehydrogenase, xanthine oxidoreductase, and myeloperoxidase intentionally produce free radical by aerobic cellular respiration (Gyllenhammar 1987; Ohara et al. 1993; Griendling et al. 1994). So, these reductase enzymes act as a reducing agent for the lucigenin. The mechanism of chemiluminescence involves the reduction of lucigenin into cation radical ($LC^{\cdot +}$). This univalent radical reacts with superoxide to produce dioxethane intermediate which breaks down to generate photons by a light-emitting process.

$$LC^{2+} + e^{-} \rightarrow LC^{\cdot +}$$

$$LC^{\cdot +} + O_2^{\cdot -} \rightarrow \text{Products} + \text{Light}$$

The intensity of the emitted light is taken as an indirect measurement of the intracellular free radicals. Previous study suggests Superoxide alone will not able to reduced lucigenin and an additional reducing agent is required (Faulkner and Fridovich 1993). Lucigenin reduced in the presence of cellular reductases enzymes including xanthine oxidase and NAD(P)H-oxidase which involve in mitochondrial respiration and lead to the direct generation of free radicals (Fig. 11.11).

Fig. 11.11 Structure of Lucigenin

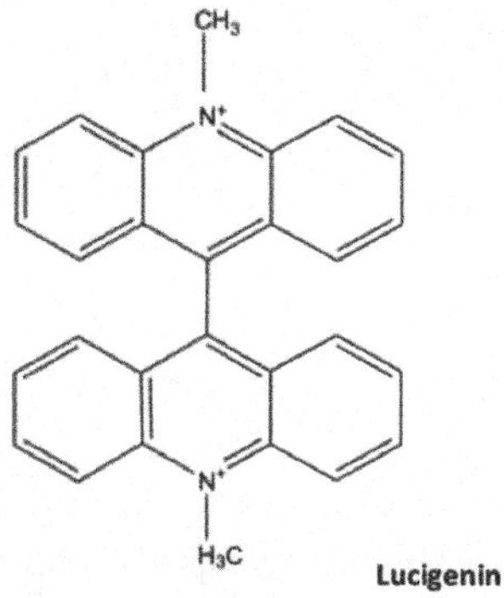

11.8 Chemical ROS Reporters

11.8.1 C11-BODIPY581/591 (Lipid Peroxidation Sensor)

Lipid peroxidation is an indirect method to measure mitochondrial redox potential in living cells. C11-BODIPY581/591 is suitable indicator for analyzing the free radical formation in vitro as well as in vivo. It is a fatty acid analog, which makes it insensitive towards $O_2^{\bullet -}$ions, nitric oxide, and hydroperoxides (Drummen et al. 2002). By the oxidation process, the red light emitting from the reduced form of the dye (595 nm) is converted in a green light-emitting oxidized form (520 nm), which leads to a change in green-to-red emission ratio (Pap et al. 1999). The quantification lipid peroxidation has been done in human skin fibroblasts by using confocal microscopy using an excitation wavelength of 488 nm (Koopman et al. 2005). Fluorescence recovery after photobleaching (FRAP) (Lippincott-Schwartz et al. 2001) analysis revealed that the laser bleach pulse (488 nm) increased with the level of the oxidized form of C11-BODIPY581/591 (Fig. 11.12).

11.8.2 HyPer

HyPer is a fluorescent ratiometric sensor for reactive species specifically for hydrogen peroxide in which cpYFP is inserted into the regulatory domain of the E. coli transcription regulator OxyR (Belousov et al. 2006). OxyR contains the H_2O_2-sensitive domain and DNA-binding domain. Upon oxidation by H_2O_2, the reduced form of OxyR is converted into an oxidized DNA-binding form. Significantly, cpYFP fluorescence was demonstrated to be very sensitive to conformational changes of an attached protein (Nagai et al. 2001). HyPer displays two excitation peaks at 420 nm (reduced form) and 500 nm (oxidized form) and a single emission peak at 516 nm. The 500/420 HyPer signal ratio has been used to measure the H_2O_2 level.

Fig. 11.12 Structure of C11-BODIPY581/591

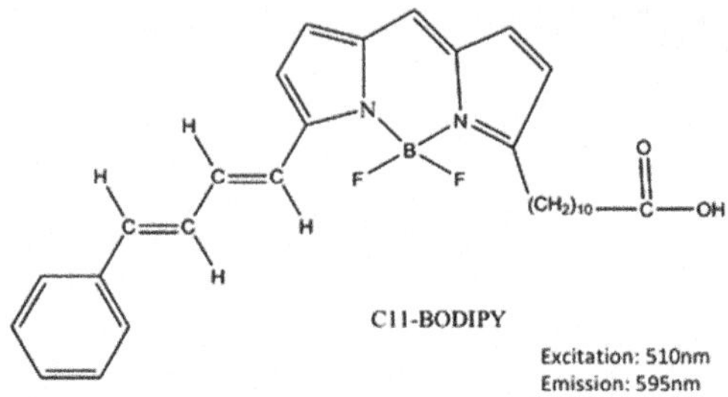

HyPer was rapidly oxidized by sub-micromolar quantity of H_2O_2in highly reducing intracellular environments (Markvicheva et al. 2009). The minimal amount of H_2O_2 has been required to induce a detectable change in HyPer fluorescence. It was demonstrated that HyPer is more specific for $H_2O_2$and did not react with other free radicals present in an intracellular environment like O_2•−, GSSG, NO_2, and ONOO− (Belousov et al. 2006).

11.9 Fibre Optics Sensor for ROS

Tan et al. (1992) firstly contributed to the development of fiber optics nanosensor for improvement of further understanding of dynamic cell capacities in vivo. Fiber optics sensors are formed by impregnation of the DCFH (2′, 7-′-dichlorodihydrofluorescein) into the DNA matrix during complexation reaction with a cationic surfactant, CTAB (Cetyltrimethylammonium bromide) in aqueous solution. For direct measurement of oxidative stress in physiologically relevant environment, a chemically incised fiber tip has synthesized by dipping it into the biomaterial (DNA-DCFH-CTAB complex) in alcohol solution (dip-coating) (Sarkar et al. 2017).

11.10 Electrochemical Sensor for Detection of ROS

A number of tools have been developed for analyzing and measurement of ROS/RNS, in which fluorescence-based probes and chemiluminescence reporters are more suitable for biological systems (Kielland et al. 2009). These approaches have some limitations due to their lack of selectivity and specificity towards free oxidant.

These electrochemical sensors are made up of using multiwall-carbon nanotube (MWCNT) composite electrodes and coating it with platinum black (Fagan-Murphy et al. 2016). These probes have been used to evaluate the levels of free oxidant were monitored from homogenates of young and old central nervous system (CNS) dissected from the pond snail, *Lymnaea stagnalis*. The normalized current responses of the probes show potential and more accurate that is required to detect the various free radicals.

11.11 Mitochondrial Targeting Sensors

Mitochondrial cellular respiration process plays an important role in free radical production within cells which also involved in regulation cell cycle and promising therapeutic targets. There are so many vectors which already developed that have ability to penetrate the mitochondrial membrane. The basic necessity to serve these functions, probes need to cross the cell membrane, phospholipid bilayer membranes of mitochondria to access intracellular organelle. Some researcher also focused on detecting reactive species in isolated mitochondria (Barja 2002; Murphy 2009), there is considerable interest in performing measurements in living cells.

These vectors exploit the higher mitochondrial potential, which help it to provide chemical opportunity to select the target organelle. These vectors mainly have two components, one is positive charges and the other is lipophilic groups (Hoyer et al. 2008; Smith et al. 2011). The lipophilic part helps in transportation across the mitochondrial membrane, whereas the positively charge anion works on the proton gradient and electrochemical potential which is generated in the mitochondrial matrix. Good ROS probes must enter into mitochondria without affect their polarization status and potential gradient. It is also possible that the loading of the sensors in mitochondria could disturb the morphology and physiology of the cell and influence the potential gradient as well as the level of free oxidant radicals.

11.11.1 Triphenylphosphonium-Based Probes

The triphenylphosphonium (TPP) group is the positively charged phosphonium compounds containing three lipophilic phenyl groups (Ross et al. 2005). Numerous sensors come into existence which have TPP group (Fig. 11.13) and allow the detection of free radical by various techniques like fluorescence intensity, mass spectroscopy, and spin trap. Mito-HE, better known as MitoSOX (Mitochondrial-targeted DHE) is widely used to solve this purpose commercially. MitoSOX is based on DHE in which the TPP group is attached to hexyl carbon chain.

MitoPY1 is also TPP-based sensors, which was used for the detection of H_2O_2 (Dickinson and Chang 2008). WhenH_2O_2react with protecting boranate moiety, this

Fig. 11.13 Structure of TPP-based probe, MITO-HE

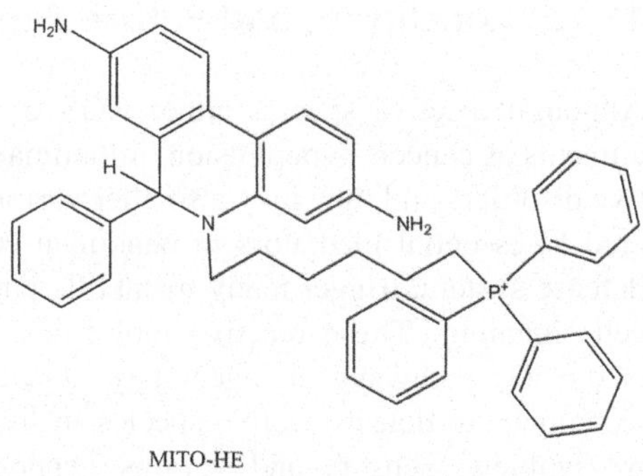

will allow emission of light through eliminating the boronate moiety. Cochemé et al. 2012 was recently developed a new sensor MitoB that is detected by Mass spectroscopy.

11.11.2 Lipophilic Cationic Compounds

Rhodamine and rosamine sensors allocate the property of encompassing a cationic functionality in the nonpolar framework. Accumulation of rhodamine 123 (RH123) in mitochondria leads to their oxidation and creates cationicrhodamine123 which exhibit green fluorescence. RH123 and Mito Tracker series probes are mainly used for mitochondrial imaging and evaluation of mitochondrial potential. These parental compounds have been used for the development of various fluorescent-based probes for the detection of ROS. It not only reacts with various reactive species but also with other oxidizing species present in the intracellular environment (Crow 1997). So, DHR cannot efficiently quantify of ROS level in the mitochondria. Koide et al., in 2007 has been developed rhodamine like sensors (MitoHR and MitoAR).

11.11.3 Peptide Delivery Agents

Peptides contain both natural and non-natural occurring amino acids, which are designed by modulating lipophilic and positively charged moieties. The Kelley laboratory-created mitochondrial penetrating peptides (MPP) which are synthesized combining synthetic and natural residues that can be cationic (e.g., arginine) or hydrophobic (e.g., cyclohexylalanine) (Horton et al. 2008; Yousif et al. 2009). In 2006, Szeto et al. designed mitochondrial-targeted peptides with vital antioxidant activity. The solid-phase synthesis of such peptides is relatively straightforward, and sensors can be added through peptide bond formation. Change in mitochondrial membrane potential or mitochondria numbers are measured by the fluorescent lifetime method so free radicals will not influence the quenching and their quantification will not be affected.

11.12 Conclusion and Future Perspective

Although reactive species either ROS or RNS have been implicated in various ailments as cancer, hypertension, inflammatory diseases, IR injuries neurodegenerative disorders and they may also alter various metabolic pathways. ROS are considered as essential mediators in beneficial cellular processes like act as pathogenic defense system, trigger many metabolic pathways, and act as messenger in normal cell signaling. These reactive molecules are produced by a number of various enzymatic pathways endogenously as well as exogenously. There are various techniques to detect reactive species in living tissues. The field of ROS detection by synthetic sensors and various genetically encoded fluorescent probe-based sensors continues to expand bringing, need an additional research to the development of robust techniques, which have high specificity and selectivity for them. An ideal ROS sensor must be more accurate, non-invasive in nature, explore a full range of concentration, applicable in vitro and in vivo, and also compatible with in vivo imaging. It should not be affected by pH, salt concentration, redox equilibrium of cell, and other environmental factors.

Acknowledgments The first author (NA) is thankful to the Department of Biotechnology (DBT), Govt. of India for grants Senior Research Fellowship.

References

Allen RC (1986) Phagocytic leukocyte oxygenation activities and chemiluminescence: a kinetic approach to analysis. Method Enzymol 133:449–493

Arnhold J (2004) Properties, functions, and secretion of human myeloperoxidase. Biochemistry (Mosc) 69(1):4–9

Barja G (2002) The quantitative measurement of H_2O_2Generation in isolated mitochondria. J Bioenerg Biomembr 34:227–233

Beatty K, Williams M, Carlson B, Swarts B, Warren R, Van Helden P (2013) Sulfatase-activated fluorophores for rapid discrimination of mycobacterial species and strains. Proc Natl Acad Sci U S A 110:12911–12916

Belousov VV, Fradkov AF, Lukyanov KA, Staroverov DB, Shakhbazov KS, Terskikh AV, Lukyanov S (2006) Genetically encoded fluorescent indicator for intracellular hydrogen peroxide. Nat Methods 3:281–286

Chen L, Sun W, Li J, Liu Z, Ma Z, Zhang W (2013) The first ratiometric fluorescent probes for aminopeptidase N cell imaging. Org Biomol Chem 11:378–382

Cochemé HM, Logan A, Prime TA, Abakumova I, Quin C, McQuaker SJ, Patel JV, Fearnley IM, James AM, Porteous CM, Smith RAJ, Hartley RC, Partridge Land Murphy MP (2012) Using the mitochondria-targeted ratiometric mass spectrometry probe Mito B to measure H_2O_2 in living drosophila. Nat Protoc 7:946–958

Crow JP (1997) Dichlorodihydrofluorescein and dihydrorhodamine 123 are sensitive indicators of peroxynitrite in vitro: implications for intracellular measurement of reactive nitrogen and oxygen species. Nitric Oxide 1:145–157

Dickinson BC, Chang CJ (2008) A targetable fluorescent probe for imaging hydrogen peroxide in the mitochondria of living cells. J Am Chem Soc 130:9638–9639

Drummen GP, Van- Liebergen LC, den Kamp JAO, Post JA (2002) C11-BODIPY 581/591, an oxidation-sensitive fluorescent lipid peroxidation probe:(micro) spectroscopic characterization and validation of methodology. Free Radic Biol Med 33:473–490

Fagan-Murphy A, Hachoumi L, Yeoman MS, Patel BA (2016) Electrochemical sensor for the detection of multiple reactive oxygen and nitrogen species from ageing central nervous system homogenates. Mech Ageing Dev 160:28–31

Faulkner K, Fridovich I (1993) Luminol and lucigenin as detectors for O_2⍰–. Free Radic Biol Med 15:447–451

Förster T (1948) Zwischenmolekulare energiewanderung und fluoreszenz. Annalen der Physic 437:55–75

Frank J, Pompella A, Biesalski HK (2000) Histochemical visualization of oxidant stress. Free Radic Biol Med 29:1096–1105

Gajovic-Eichelmann N, Ehrentreich-Förster E, Bier FF (2003) Directed immobilization of nucleic acids at ultramicroelectrodes using a novel electro- deposited polymer. Biosens Bioelectron 19:417–422

Gielis JF, Lin JY, Wingler K (2011) Pathogenetic role of eNOS uncoupling in cardiopulmonary disorders. Free Radic Biol Med 50:765–776

Gillissen A, Nowak D (1998) Characterization of N-acetylcysteine and ambroxol in anti-oxidant therapy. Respir Med 92:609–623

Gomes A, Eduarda F, Lima JLFC (2005) Fluorescence probes used for detection of reactive oxygen species. J Biochem Biophys Methods 65:45–80

Griendling KK, Minieri CA, Ollerenshaw JD, Alexander RW (1994) Angiotensin II stimulates NADH and NADPH oxidase activity in cultured vascular smooth muscle cells. Circ Res 74:1141–1148

Gyllenhammar H (1987) Lucigenin chemiluminescence in the assessment of neutrophil superoxide production. J Immunol Methods 97:209–213

Horton KL, Stewart KM, Fonseca SB, Guo Q, Kelley SO (2008) Mitochondria-penetrating peptides. Chem Biol 15:375–382

Hoye AT, Davoren JE, Wipf P, Fink MP, Kagan VE (2008) Targeting mitochondria. Acc Chem Res 41:87–97

Ischiropoulos H, Gow A, Thom SR, Kooy NW, Royall JA, Crow JP (1999) Detection of reactive nitrogen species using 2, 7-dichlorodihydrfluorescein and dihydrorhodamine 123. Methodenzymol 301:367–373

Kielland A, Blom T, Nandakumar KS, Holmdahl R, Blomhoff R, Carlsen H (2009) In vivo imaging of reactive oxygen and nitrogen species in inflammation using the luminescent probe L-012. Free Radic Biol Med 47:760–766

Koide Y, Kawaguchi M, Urano Y, Hanaoka K, Komatsu T, Abo M (2012) A reversible near-infrared fluorescence probe for reactive oxygen species based on Te-rhodamine. Chem Commun (Camb) 48:3091–3093

Koide Y, Urano Y, Kenmoku S, Kojima H, Nagano T (2007) Design and synthesis of fluorescent probes for selective detection of highly reactive oxygen species in mitochondria of living cells. J Am Chem Soc 129:10324–10325

Komatsu H, Harada H, Tanabe K, Hiraoka M, Nishimoto S (2010) Indolequinone-rhodol conjugate as a fluorescent probe for hypoxic cells: enzymatic activation and fluorescence properties. Med Chem Comm 1:50

Koopman WJ, Verkaart S, van Emst-de Vries SE, Grefte S, Smeitink JA, Willems PH (2006) Simultaneous quantification of oxidative stress and cell spreading using 5-(and-6)-chloromethyl-2′, 7′-dichlorofluorescein. Cytometry A 69:1184–1192

Koopman WJ, Verkaart S, Visch HJ, van der Westhuizen FH, Murphy MP, van den Heuvel LW, Willems PH (2005) Inhibition of complex I of the electron transport chain causes O_2–-·-mediated mitochondrial outgrowth. AmJ Physiol Cell Physiol 288:C1440–C1450

Li X, Zhang H, Xie Y, Hu Y, Sun H, Zhu Q (2014) Fluorescent probes for detecting monoamine oxidase activity and cell imaging. Org Biomol Chem 12:2033–2036

Lippincott-Schwartz J, Snapp E, Kenworthy A (2001) Studying protein dynamics in living cells. Nat Rev Mol Cell Biol 2:444–456

Lovell JF, Chen J, Jarvi MT, Cao WG, Allen AD, Liu Y, Zheng G (2009) FRET quenching of photosensitizer singlet oxygen generation. J Phys Chem 113:3203–3211

Markvicheva KN, Bogdanova EA, Staroverov DB, Lukyanov S, Belousov VV (2009) Imaging of intracellular hydrogen peroxide production with Hyper upon stimulation of Hela cells with Egf. In: Redox-mediated signal transduction. Methods and protocol. Humana Press, New York, pp 76–83

Miller E, Bian S, Chang C (2007) A fluorescent sensor for imaging reversible redox cycles in living cells. J Am Chem Soc 129:3458–3459

Mohanty JG, Jaffe JS, Shulman ES, Raible DG (1997) A highly sensitive fluorescent micro-assay of H_2O_2 release from activated human leukocytes using a dihydroxyphenoxazine derivate. J Immunol Methods 202:133–141

Murphy MP (2009) How mitochondria produce reactive oxygen species. Biochem J 417:1–13

Nagai T, Sawano A, Park ES, Miyawaki A (2001) Circularly permuted green fluorescent proteins engineered to sense Ca_2^+. Proc Natl Acad Sci U S A 98:3197–3202

Ohara Y, Peterson TE, Harrison DG (1993) Hypercholesterolemia increases endothelial superoxide anion production. J Clin Investig 91:2546

Pap EHW, Drummen GPC, Winter VJ, Kooij TWA, Rijken P, Wirtz KWA, Post JA (1999) Ratio-fluorescence microscopy of lipid oxidation in living cells using C11-BODIPY581/591. FEBS Lett 453:278–282

Park HS, Park D, Bae YS (2006) Molecular interaction of NADPH oxidase 1 with betaPix and Nox organizer 1. Biochem Bioph Res Co 339:985–990

Robinson KM, Janes MS, Pehar M, Monette JS, Ross MF, Hagen TM, Murphy MP, Beckman JS (2006) Selective fluorescent imaging of superoxide in vivo using ethidium- based probes. Proc Natl Acad Sci U S A 103:15038–15043

Ross MF, Kelso GF, Blaikie FH, James AM, Cocheme HM, Filipovska A, Da Ros T, Hurd TR, Smith RAJ, Murphy MP (2005) Lipophilic triphenylphosphonium cations as tools in mitochondrial bioenergetics and free radical biology. Biochem Mosc 70(2):222–230

Royall JA, Ischiropoulos H (1993) Evaluation of 2′,7′-dichlorofluorescin and dihydrorhodamine 123 as fluorescent probes for intracellular H_2O_2 in cultured endothelial cells. Arch Biochem Biophys 302:348–355

Sarkar PK, Halder A, Adhikari A, Polley N, Darbar S, Lemmens P, Pal SK (2017) DNA-based fiber optic sensor for direct in-vivo measurement of oxidative stress. Sensor Actuat B-Chem 255:2194–2202

Singh R, Singh S, Parihar P, Mishra RK, Tripathi DK, Singh VP, Chauhan DK, Tan W, Shi ZY, Smith S, Birnbaum D, Kopelman R (1992) Submicrometer intracellular chemical optical fibre sensors. Science 258:778–781

Smith RAJ, Hartley RC, Murphy MP (2011) Mitochondria-targeted small molecule therapeutics and probes. Antioxid Redox Signal 15:3021–3038

Sundaresan M, Yu ZX, Ferrans VJ, Irani K, Finkel T (1995) Requirement for generation of H_2O_2 for platelet-derived growth factor signal transduction. Science 270(5234):296

Suzuki YZ, Forman HJ, Sevanian A (1997) Oxidants as stimulators of signal transduction. Free Radic Biol Med 22:269–285

Szeto HH (2006) Cell-permeable, mitochondrial-targeted, peptide antioxidants. AAPS J 8:E277–E283

Tan W, Shi ZY, Smith S, Birnbaum D, Kopelman R (1992) Submicrometer intracellular chemical optical fibre sensors. Science 258:778–781

Waris G, Ahsan H (2006) Reactive oxygen species: role in the development of cancer and various chronic conditions. J Carcinog 5:14

White KA, Marletta MA (1992) Nitric oxide synthase is a cytochrome P-450 type hemoprotein. Biochemistry 31:6627–6631

Yousif LF, Stewart KM, Horton KL, Kelley SO (2009) Mitochondria-penetrating peptides: sequence effects and model cargo transport. Chem Bio Chem 10:2081–2088

Zhao H, Joseph J, Fales HM, Sokoloski EA, Levine RL, Vasquez-Vivar J, Zuo L, Otenbaker NP, Rose BA, Salisbury KS (2013) Molecular mechanisms of reactive oxygen species- related pulmonary inflammation and asthma. Mol Immunol 56:57–63

Zhou M, Diwu Z, Panchuk-Voloshina N, Haugland RP (1997) A stable nonfluorescent derivative of resorufin for the fluorometric determination of trace hydrogen peroxide: applications in detecting the activity of phagocyte NADPH oxidase and other oxidases. Anal Biochem 253:162–168

Zuo L, Otenbaker NP, Rose BA, Salisbury KS (2013) Molecular mechanisms of reactive oxygen species-related pulmonary inflammation and asthma. Mol Immunol 56:57–63

The manufacturer's authorised representative in the EU is Springer Nature Customer Service Centre GmbH, Europaplatz 3, 69115 Heidelberg, Germany. If you have any concerns regarding our products, please contact ProductSafety@springernature.com

Printed and bound by CPI Group (UK) Ltd, Croydon, CR0 4YY
30/09/2026
02225291-0004